Unive

Roger Mansuy · Marc Yor

Aspects of
Brownian Motion

 Springer

Roger Mansuy
21, Boulevard Carnot
92340 Bourg-la-Reine
France

Marc Yor
Université Paris VI
Laboratoire de Probabilités et
Modèles Aléatoires
4, place Jussieu
75252 Paris Cedex 5
France
deaproba@proba.jussieu.fr

An earlier version of this book was published by Birkhäuser, Basel
as Yor, Marc: Some aspects of Brownian Motion, Part I, 1992, and
Yor, Marc: Some aspects of Brownian Motion. Part II, 1997.

ISBN 978-3-540-22347-4   ISBN 978-3-540-49966-4 (eBook)

Library of Congress Control Number: 2008930798

Mathematics Subject Classification (2000): 60-02, 60-01, 60J65, 60E05

© 2008 Springer-Verlag Berlin Heidelberg

Cover design: WMX Design GmbH, Heidelberg
The cover illustration is based on a simulation of BESQ processes provided by C. Umansky.

Printed on acid-free paper

9 8 7 6 5 4 3 2 1

springer.com

Introduction

This volume is the result of our efforts to update the eleven first chapters of the two previously published ETH Zürich Lecture Notes by the second author: *Some Aspects of Brownian Motion, Part I (1992); Part II (1997)*. The original volumes have been out of print since, roughly, the year 2000. We have already updated the remaining chapters of Part II in:
Random Times and Enlargements of Filtrations in a Brownian Setting, Lecture Notes in Maths, n°1873, Springer (2006).
Coming back to the present volume, we modified quite minimally the old eleven first chapters, essentially by completing the Bibliography. Here is a detailed description of these eleven chapters; each of them is devoted to the study of some particular class of Brownian functionals; these classes appear in increasing order of complexity.

In Chapter 1, various results about certain Gaussian subspaces of the Gaussian space generated by a one-dimensional Brownian motion are obtained; the derivation of these results is elementary in that it uses essentially Hilbert spaces isomorphisms between certain Gaussian spaces and some L^2 spaces of deterministic functions.

In Chapter 2, several results about Brownian quadratic functionals are obtained, with some particular emphasis on a change of probability method, which enables to obtain a number of variants of Lévy's formula for the stochastic area of Brownian motion.

In Chapter 3, Ray-Knight theorems on Brownian local times are recalled and extended; the processes which appear there are squares of Bessel processes, which links naturally chapter 3 with the study of Brownian quadratic functionals made in chapter 2; in the second half of chapter 3, some relations with Bessel meanders and bridges are discussed.

In Chapter 4, the relation between squares of Bessel processes and Brownian local times is further exploited, in order to explain and extend the Ciesielski-Taylor identities.

In Chapters 5 and 7, a number of results about Brownian windings are established; exact distributional computations are made in chapter 5, whereas asymptotic studies are presented in chapter 7.

Chapter 6 is devoted to the study of the integral, on a time interval, of the exponential of a Brownian motion with drift; this study is important in mathematical finance.

In Chapters 8 and 9, some extensions of Paul Lévy's arc sine law for Brownian motion are discussed, with particular emphasis on the time spent by Brownian motion below a multiple of its one-sided supremum.

Principal values of Brownian and Bessel local times - in particular their Hilbert transforms - are discussed in **Chapter 10**. Such principal values occur naturally in the Dirichlet decomposition of Bessel processes with dimension smaller than 1, as well as when considering certain signed measures which are absolutely continuous with respect to the Wiener measure.

The Riemann zeta function and Jacobi theta functions are shown, in **Chapter 11**, to be somewhat related with the Itô measure of Brownian excursions. Some generalizations to Bessel processes are also presented.

We are well aware that this particular selection of certain aspects of Brownian motion is, at the same time, quite incomplete and arbitrary, but in the defense of our choice, let us say that:

a. We feel some confidence with these particular aspects...

b. Some other aspects are excellently treated in a number of lecture notes and books, the references of which are gathered at the end of this volume.

c. Between 2004 and 2006, we had undertaken an ambitious updating of the same ETH Lecture Notes , but were unable to complete this more demanding task. The interested reader may consult online (http://roger.mansuy.free.fr/Aspects/Aspects_ references.html) the extensive Bibliography we had gathered for this purpose.

Many thanks to Kathleen Qechar for juggling with the different versions, macros, and so on...

Brannay, May 4th, 2008.

Keywords, chapter by chapter

Chapter 1: Gaussian space, first Wiener chaos, filtration of Brownian bridges, ergodic property, space-time harmonic functions.

Chapter 2: Quadratic functionals, Lévy's area formula, Ornstein-Uhlenbeck process, Fubini-Wiener integration by parts.

Chapter 3: Ray-Knight theorems, transfer principle, additivity property, Lévy-Khintchine representation, generalized meanders, Bessel bridges.

Chapter 4: Ciesielski-Taylor (: CT) identities, Biane's extensions.

Chapter 5: Winding number, Hartman-Watson distribution, Brownian lace.

Chapter 6: Asian options, Confluent hypergeometric functions, beta and gamma variables.

Chapter 7: Kallianpur-Robbins ergodic theorem, Spitzer's theorem, Gauss linking number.

Chapter 8: P. Lévy's arc sine law, F. Petit's extensions, Walsh's Brownian motion, Excursion theory Master formulae, Feynman-Kac formula.

Chapter 9: Local time perturbation of Brownian motion, Bismut's identity, Knight's ratio formula.

Chapter 10: Hilbert transform, principal values, Yamada's formulae, Dirichlet processes, Bertoin's excursion theory for BES(d).

Chapter 11: Riemann Zeta function, Jacobi theta function, Convolution of Hitting times, Chung's identity.

Contents

6   On some exponential functionals of Brownian motion and the problem of Asian options 79

7   Some asymptotic laws for multidimensional BM 101

8   Some extensions of Paul Lévy's arc sine law for BM 115

Chapter 1

The Gaussian space of BM

In this Chapter, a number of linear transformations of the Gaussian space associated to a linear Brownian motion $(B_t, t \geq 0)$ are studied. Recall that this Gaussian space is precisely equal to the first Wiener chaos of B, that is:

$$\Gamma(B) \stackrel{\text{def}}{=} \left\{ B^f \equiv \int_0^\infty f(s)dB_s \ , \quad f \in L^2(\mathbb{R}_+, ds) \right\}$$

In fact, the properties of the transformations being studied may be deduced from corresponding properties of associated transformations of $L^2(\mathbb{R}_+, ds)$, thanks to the Hilbert spaces isomorphism:

$$B^f \leftrightarrow f$$

between $\Gamma(B)$ and $L^2(\mathbb{R}_+, ds)$, which is expressed by the identity:

$$E\left[(B^f)^2\right] = \int_0^\infty dt \ f^2(t) \tag{1.1}$$

This chapter may be considered as a warm-up, and is intended to show that some interesting properties of Brownian motion may be deduced easily from the covariance identity (1.1).

1.1 A realization of Brownian bridges

Let $(B_u, u \geq 0)$ be a 1-dimensional BM, starting from 0. Fix $t > 0$ for one moment, and remark that, for $u \leq t$:

$$B_u = \frac{u}{t}B_t + \left(B_u - \frac{u}{t}B_t\right)$$

is the orthogonal decomposition of the gaussian variable B_u with respect to B_t.

Hence, since $(B_u, u \geq 0)$ is a Gaussian process, the process $\left(B_u - \frac{u}{t}B_t, u \leq t\right)$ is independent of the variable B_t.

Let now $\Omega_{(t)}^* \equiv C\left([0,t]; \mathbb{R}\right)$ be the space of continuous functions $\omega : [0,t] \to \mathbb{R}$; on $\Omega_{(t)}^*$, denote $X_u(\omega) = \omega(u)$, $u \leq t$, and $\mathcal{F}_u = \sigma\{X_s, s \leq u\}$. $\mathcal{F}_t$ is also the Borel σ-field when $\Omega_{(t)}^*$ is endowed with the topology of uniform convergence.

For any $x \in \mathbb{R}$, we define $P_x^{(t)}$ as the distribution on $(\Omega_{(t)}^*, \mathcal{F}_t)$ of the process:

$$\left(\frac{ux}{t} + B_u - \frac{u}{t}B_t; u \leq t\right) \quad .$$

Clearly, the family $(P_x^{(t)}; x \in \mathbb{R})$ is weakly continuous, and, by construction, it satisfies:

$$E\left[F(B_u, u \leq t) \mid B_t = x\right] = E_x^{(t)}\left[F(X_u, u \leq t)\right] \quad dx \quad \text{a.e. },$$

for every $(\mathcal{F}_t)$ measurable, bounded functional F. Hence, there is no ambiguity in defining, for any $x \in \mathbb{R}$, $P_x^{(t)}$ as the law of the Brownian bridge, of duration t, starting at 0, and ending at x.

We shall call $P_0^{(t)}$ the law of the standard Brownian bridge of duration t. Hence, a realization of this bridge is:

$$\left(B_u - \frac{u}{t}B_t; u \leq t\right) \quad .$$

1.2 The filtration of Brownian bridges

If G is a subset of the Gaussian space generated by $(B_u, u \geq 0)$, we denote by $\Gamma(G)$ the Gaussian space generated by G, and we use the script letter $\mathcal{G}$ for the σ-field $\sigma(G)$.

We now define $\Gamma_t = \Gamma(G_t)$, where $G_t = \left\{ B_u - \frac{u}{t} B_t; u \leq t \right\}$ and $\mathcal{G}_t = \sigma(G_t)$. It is immediate that Γ_t is the orthogonal of $\Gamma(B_t)$ in $\Gamma(B_u, u \leq t)$, that is, we have:

$$\Gamma(B_u, u \leq t) = \Gamma_t \oplus \Gamma(B_t) \ .$$

Remark that $\{\Gamma_t, t \geq 0\}$ is an increasing family, since, for $u \leq t \leq t + h$:

$$B_u - \frac{u}{t} B_t = \left(B_u - \frac{u}{t+h} B_{t+h} \right) - \frac{u}{t} \left(B_t - \frac{t}{t+h} B_{t+h} \right) \ ,$$

and that, moreover: $\Gamma_\infty \overset{\text{def}}{=} \lim_{t \uparrow \infty} \uparrow \Gamma_t \equiv \Gamma(B_u, u \geq 0)$, since:

$$B_u = \text{a.s.} \lim_{t \to \infty} \left(B_u - \frac{u}{t} B_t \right)$$

Hence, $(\mathcal{G}_t, t \geq 0)$ is a subfiltration of $(\mathcal{B}_t \equiv \sigma(B_u, u \leq t), t \geq 0)$, and $\mathcal{G}_\infty = \mathcal{B}_\infty$. Here are some more precisions about $(\mathcal{G}_t, t \geq 0)$.

Theorem 1.1 *1) For any $t > 0$, we have:*

$$\Gamma_t = \left\{ \int_0^t f(u)dB_u; f \in L^2\left([0,t], du\right), \quad \text{and} \quad \int_0^t du\, f(u) = 0 \right\}$$

2) For any $t > 0$, the process:

$$\gamma_u^{(t)} = B_u - \int_0^u ds \frac{B_t - B_s}{t - s}, \quad u \leq t \ ,$$

is a Brownian motion, which is independent of the variable B_t. Moreover, we have: $\Gamma_t = \Gamma(\gamma_u^{(t)}, u \leq t)$

3) *The process:* $\beta_t = B_t - \int_0^t \dfrac{ds}{s} B_s$, $t \geq 0$, *is a Brownian motion, and we have:*

$$\Gamma_t = \Gamma(\beta_s, s \leq t) \ .$$

Consequently, $(\mathcal{G}_t, t \geq 0)$ *is the natural filtration of the Brownian motion* $(\beta_t, t \geq 0)$.

PROOF:

1) The first assertion of the Theorem follows immediately from the Hilbert spaces isomorphism between $L^2([0,t], du)$ and G_t, which transfers a function f into $\int_0^t f(u)dB_u$

2) Before we prove precisely the second and third assertions of the Theorem, it is worth explaining how the processes $(\gamma_u^{(t)}, u \leq t)$ and $(\beta_t, t \geq 0)$ arise naturally. It is not difficult to show that $(\gamma_u^{(t)}, u \leq t)$ is the martingale part in the canonical decomposition of $(B_u, u \leq t)$ as a semimartingale in the filtration $\left\{ \mathcal{B}_u^{(t)} \equiv \mathcal{B}_u \vee \sigma(B_t); u \leq t \right\}$, whereas the idea of considering $(\beta_u, u \geq 0)$ occured by looking at the Brownian motion $(\gamma_u^{(t)}, u \leq t)$, reversed from time t, that is:

$$\gamma_t^{(t)} - \gamma_{t-u}^{(t)} = (B_t - B_{t-u}) - \int_0^u ds \frac{B_t - B_{t-s}}{s} \ .$$

3) Now, let $(Z_u, u \leq t)$ be a family of Gaussian variables which belong to Γ_t; in order to show that $\Gamma_t = \Gamma(Z_u, u \leq t)$, it suffices, using the first assertion of the theorem, to prove that the only functions $f \in L^2([0,t], du)$ such that

$$E\left[Z_u \left(\int_0^t f(v)dB_v \right) \right] = 0 \ , \quad \text{for every } u \leq t \tag{1.2}$$

are the constants.

When we apply this remark to $Z_u = \gamma_u^{(t)}, u \leq t$, we find that f satisfies (1.2) if and only if

$$\int_0^u dv\, f(v) - \int_0^u ds \frac{1}{(t-s)} \int_s^t dv\, f(v) = 0\ , \qquad \text{for every } u \le t,$$

hence:

$$f(v) = \frac{1}{t-v} \int_v^t du\, f(u)\ , \qquad dv \text{ a.s.},$$

from which we now easily conclude that $f(v) = c$, dv a.s., for some constant c. A similar discussion applies with $Z_u = \beta_u$, $u \le t$. $\square$

Exercise 1.1:

Let $f : \mathbb{R}_+ \to \mathbb{R}$ be an absolutely continuous function which satisfies:
$$f(0) = 0, \text{ and for } t > 0,\ f(t) \neq 0, \text{ and } \int_0^t \frac{du}{|f(u)|} \left(\int_0^u (f'(s))^2 ds \right)^{1/2} < \infty$$

1. Show that the process:

$$Y_t^{(f)} = B_t - \int_0^t \frac{du}{f(u)} \left(\int_0^u f'(s) dB_s \right)\ , \qquad t \ge 0\ ,$$

admits $(\mathcal{G}_t)$ as its natural filtration.

2. Show that the canonical decomposition of $(Y_t^{(f)}, t \ge 0)$ in its natural filtration $(\mathcal{G}_t)$ is:

$$Y_t^{(f)} = \beta_t + \int_0^t \frac{du}{f(u)} \left(\int_0^u \left(\frac{f(s)}{s} - f'(s) \right) d\beta_s \right)\ .$$

1.3 An ergodic property

We may translate the third statement of Theorem 1.1 by saying that, if $(X_t, t \ge 0)$ denotes the process of coordinates on the canonical space $\Omega^* \equiv \Omega^*_{(\infty)} \equiv C([0, \infty), \mathbb{R})$, then the transformation T defined by:

$$T(X)_t = X_t - \int_0^t \frac{ds}{s} X_s \quad (t \geq 0)$$

leaves the Wiener measure W invariant.

Theorem 1.2 *For any $t > 0$, $\bigcap_n (T^n)^{-1}(\mathcal{F}_t)$ is W-trivial. Moreover, for any $n \in \mathbb{N}$, we have: $(T^n)^{-1}(\mathcal{F}_\infty) = \mathcal{F}_\infty$, W a.s. (in the language of ergodic theory, T is a K-automorphism). Consequently, the transformation T on $(\Omega^*, \mathcal{F}_\infty, W)$ is strongly mixing and, a fortiori, ergodic.*

PROOF:

a) The third statement follows classically from the two first ones.

b) We already remarked that $T^{-1}(\mathcal{F}_\infty) = \mathcal{F}_\infty$, W a.s., since $\mathcal{G}_\infty = \mathcal{B}_\infty$, which proves the second statement.

c) The first statement shall be proved later on as a consequence of the next Proposition 1.1. □

To state simply the next Proposition, we need to recall the definition of the classical Laguerre polynomials:

$$L_n(x) = \sum_{k=0}^n \binom{n}{k} \frac{1}{k!} (-x)^k , \quad n \in \mathbb{N} ,$$

is the sequence of orthonormal polynomials for the measure $e^{-x} dx$ on $\mathbb{R}_+$ which is obtained from $(1, x, x^2, \ldots, x^n, \ldots)$ by the Gram-Schmidt procedure.

Proposition 1.1 *Let $(X_t)_{t \leq 1}$ be a real-valued BM, starting from 0. Define $\gamma_n = T^n(X)_1$. Then, we have:*

$$\gamma_n = \int_0^1 dX_s L_n \left(\log \frac{1}{s} \right) .$$

$(\gamma_n, n \in \mathbb{N})$ is a sequence of independent centered Gaussian variables, with variance 1, from which $(X_t, t \leq 1)$ may be represented as:

$$X_t = \sum_{n \in \mathbb{N}} \lambda_n \left(\log \frac{1}{t} \right) \gamma_n, \quad \text{where } \lambda_n(a) = \int_0^a dx \, e^{-x} L_n(x)$$

PROOF: The expression of γ_n as a Wiener integral involving L_n is obtained by iteration of the transformation T.

The identity: $E[\gamma_n \gamma_m] = \delta_{nm}$ then appears as a consequence of the fact that the sequence $\{L_n, n \in \mathbb{N}\}$ constitutes an orthonormal basis of $L^2(\mathbb{R}_+, e^{-x}dx)$.

Indeed, we have:

$$E[\gamma_n \gamma_m] = \int_0^1 ds \, L_n \left(\log \frac{1}{s} \right) L_m \left(\log \frac{1}{s} \right) = \int_0^\infty dx \, e^{-x} L_n(x) L_m(x) = \delta_{nm} \quad .$$

More generally, the application:

$$(f(x), x > 0) \longrightarrow \left(f \left(\log \frac{1}{s} \right), 0 < s < 1 \right)$$

is an isomorphism of Hilbert spaces between $L^2(e^{-x}dx; \mathbb{R}_+)$ and $L^2(ds; [0,1])$, and the development of $(X_t)_{t \leq 1}$ along the (γ_n) sequence corresponds to the development of $1_{[0,t]}(s)$ along the basis $\left(L_n \left(\log \frac{1}{s} \right) \right)_{n \in \mathbb{N}}$. $\square$

1.4 A relationship with space-time harmonic functions

In this paragraph, we are interested in a question which in some sense is dual to the study of the transformation T which we considered above. More precisely, we wish to give a description of the set $\mathcal{J}$ of all probabilities P on $(\Omega^*, \mathcal{F}_\infty)$ such that:

i) $\left(\tilde{X}_t \equiv X_t - \int_0^t \frac{ds}{s} X_s ; t \geq 0 \right)$ is a real valued BM; here, we only assume

that the integral $\int_0^t \frac{ds}{s} X_s \equiv$ a.s. $\lim_{\varepsilon \to 0} \int_\varepsilon^t \frac{ds}{s} X_s$ exists a.s., but we do not

assume a priori that is converges absolutely.

ii) for every $t \geq 0$, the variable X_t is P independent of $(\tilde{X}_s, s \leq t)$.

We obtain the following characterization of the elements of $\mathcal{J}$.

Theorem 1.3 *Let W denote the Wiener measure on $(\Omega^*, \mathcal{F}_\infty)$ (W is the law of the real valued Brownian motion B starting from 0).*

Let P be a probability on $(\Omega^*, \mathcal{F}_\infty)$.

The three following properties are equivalent:

1) $P \in \mathcal{J}$.

2) P is the law of $(B_t + tY, t \geq 0)$, where Y is a r.v. which is independent of $(B_t, t \geq 0)$;

3) there exists a function $h : \mathbb{R}_+ \times \mathbb{R} \to \mathbb{R}_+$, which is space-time harmonic, that is: such that $(h(t, X_t), t \geq 0)$ is a $(W, \mathcal{F}_t)$ martingale, with expectation 1, and $P = W^h$, where W^h is the probability on $(\Omega^*, \mathcal{F}_\infty)$ defined by:

$$W^h\big|_{\mathcal{F}_t} = h(t, X_t) \cdot W\big|_{\mathcal{F}_t} \ .$$

We first describe all solutions of the equation

$(*)$ $\qquad\qquad\qquad X_t = \beta_t + \int_0^t \frac{ds}{s} X_s \ ,$

where $(\beta_t, t \geq 0)$ is a real-valued BM, starting from 0.

Lemma 1.1 *(X_t) is a solution of $(*)$ iff there exists a r.v. Y such that:*

$$X_t = t \left(Y - \int_t^\infty \frac{d\beta_u}{u} \right) .$$

PROOF: From Itô's formula, we have, for $0 < s < t$:

$$\frac{1}{t} X_t = \frac{1}{s} X_s + \int_s^t \frac{d\beta_u}{u} .$$

As $t \to \infty$, the right-hand side converges, hence, so does the left-hand side; we call Y the limit of $\dfrac{X_t}{t}$, as $t \to \infty$; we have

$$\frac{1}{s} X_s = Y - \int_s^\infty \frac{d\beta_u}{u} .$$

□

We may now give a proof of Theorem 1.3; the rationale of the proof shall be: 1)⇒2)⇒3)⇒1).

1)⇒2): from Lemma 1.1, we have: $\dfrac{X_t}{t} = Y - \displaystyle\int_t^\infty \frac{d\tilde{X}_u}{u}$, and we now remark that

$$B_t = -t \int_t^\infty \frac{d\tilde{X}_u}{u} , \quad t \geq 0, \quad \text{is a } BM. \tag{1.3}$$

Hence, it remains to show that Y is independent from B; in fact, we have: $\sigma\{B_u, u \geq 0\} = \sigma\{\tilde{X}_u, u \geq 0\}$, up to negligible sets, since, from (1.3), it follows that:

$$d\left(\frac{B_t}{t} \right) = \frac{d\tilde{X}_t}{t} .$$

However, from our hypothesis, X_t is independent of $\tilde{X}_u, u \leq t$, so that $Y \equiv \lim_{t \to \infty} \left(\dfrac{X_t}{t} \right)$ is independent of $(\tilde{X}_u, u \geq 0)$.

2)⇒3): We condition with respect to Y; indeed, let $\nu(dy) = P(Y \in dy)$, and define:

$$h(t,x) = \int \nu(dy) \exp\left(yx - \frac{y^2 t}{2} \right) \equiv \int \nu(dy) h_y(t,x) .$$

From Girsanov's theorem, we know that:

$$P\left\{(B_u + yu, u \geq 0) \in \Gamma\right\} = W^{h_y}(\Gamma) \ ,$$

and therefore, here, we have: $P = W^h$.

3)$\Rightarrow$1): If $P = W^h$, then we know that $(\tilde{X}_u, u \leq t)$ is independent of X_t under W, hence also under P, since the density $\left.\dfrac{dP}{dW}\right|_{\mathcal{F}_t} = h(X_t, t)$ depends only on X_t. $\square$

Exercise 1.2: Let $\lambda \in \mathbb{R}$. Define $\beta_t^{(\lambda)} = B_t - \lambda \displaystyle\int_0^t \dfrac{ds}{s} B_s$ $(t \geq 0)$.

Let $\mathcal{F}_t^{(\lambda)} = \sigma\{\beta_s^{(\lambda)}; s \leq t\}$, $t \geq 0$, be the natural filtration of $(\beta_t^{(\lambda)}, t \geq 0)$, and $(\mathcal{F}_t, t \geq 0)$ be the natural filtration of $(\beta_t^{(\lambda)}, t \geq 0)$.

1. Show that $(\mathcal{F}_t^{(\lambda)}, t \geq 0)$ is a strict subfiltration of $(\mathcal{F}_t, t \geq 0)$ if, and only if, $\lambda > \frac{1}{2}$.

2. We now assume: $\lambda > \frac{1}{2}$.
   Prove that the canonical decomposition of $(\beta_t^{(\lambda)}, t \geq 0)$ in its natural filtration $(\mathcal{F}_t^{(\lambda)}, t \geq 0)$ is:

$$\beta_t^{(\lambda)} = \gamma_t^{(\lambda)} - (1-\lambda) \int_0^t \frac{ds}{s} \gamma_s^{(\lambda)} \ , \quad t \geq 0 \ ,$$

   where $(\gamma_t^{(\lambda)}, t \geq 0)$ is a $(\mathcal{F}_t^{(\lambda)}, t \geq 0)$ Brownian motion.

3. Prove that the processes: $B, \beta^{(\lambda)}$, and $\gamma^{(\lambda)}$ satisfy the following relations:

$$d\left(\frac{B_t}{t^\lambda}\right) = \frac{d\beta_t^{(\lambda)}}{t^\lambda} \quad \text{and} \quad d\left(\frac{\gamma_t^{(\lambda)}}{t^{1-\lambda}}\right) = \frac{d\beta_t^{(\lambda)}}{t^{1-\lambda}} \ .$$

Exercise 1.3: (We use the notation introduced in the statement or the proof of Theorem 1.3).

Let Y be a real-valued r.v. which is independent of $(B_t, t \geq 0)$; let $\nu(dy) = P(Y \in dy)$ and define: $B_t^{(\nu)} = B_t + Yt$.

1. Prove that if $f : \mathbb{R} \to \mathbb{R}_+$ is a Borel function, then:

$$E\left[f(Y) \mid B_s^{(\nu)}, s \leq t\right] = \frac{\displaystyle\int \nu(dy) f(y) \exp\left(yB_t^{(\nu)} - \frac{y^2 t}{2}\right)}{\displaystyle\int \nu(dy) \exp\left(yB_t^{(\nu)} - \frac{y^2 t}{2}\right)}$$

2. With the help of the space-time harmonic function h featured in property 3) of Theorem 1.3, write down the canonical decomposition of $(B_t^{(\nu)}, t \geq 0)$ in its own filtration.

1.5 Brownian motion and Hardy's inequality in L^2

(1.5.1) The transformation T which we have been studying is closely related to the Hardy transform:

$$H : L^2\left([0,1]\right) \longrightarrow L^2\left([0,1]\right)$$

$$f \longrightarrow Hf : x \to \frac{1}{x} \int_0^x dy\, f(y)$$

We remark that the adjoint of H, which we denote by $\tilde{H}$, satisfies:

$$\tilde{H}f(x) = \int_x^1 \frac{dy}{y} f(y), \quad f \in L^2\left([0,1]\right) \quad.$$

The operator $K = H$, or $\tilde{H}$, satisfies Hardy's L^2 inequality:

$$\int_0^1 dx(Kf)^2(x) \leq 4 \int_0^1 dx\, f^2(x) \quad,$$

which may be proved by several simple methods, among which one is to consider martingales defined on $[0,1]$, fitted with Lebesgue measure, and the filtration $\{\mathcal{F}_t = \sigma(A, \text{Borel set}; A \subset [0,t]; t \leq 1\}$ (see, for example, Dellacherie-Meyer-Yor [29]). In this paragraph, we present another approach, which is clearly related to the Brownian motion (β_t) introduced in Theorem 1.1. We first remark that if, to begin with, f is bounded, we may write

$$(*) \qquad \int_0^1 f(u)dB_u = \int_0^1 f(u)d\beta_u + \int_0^1 \frac{du}{u} B_u f(u),$$

and then, we remark that

$$\int_0^1 \frac{du}{u} B_u f(u) = \int_0^1 dB_u (\tilde{H}f)(u) \ ;$$

hence, from $(*)$

$$\int_0^1 dB_u (\tilde{H}f)(u) = \int_0^1 f(u)dB_u - \int_0^1 f(u)d\beta_u \ ,$$

from which we immediately deduce Hardy's L^2 inequality.

We now go back to $(*)$ to remark that, for any $f \in L^2[0,1]$, or, in fact more generally, for any $(\mathcal{G}_u)_{u \leq 1}$ predictable process $(\varphi(u, \omega))$ such that:

$$\int_0^1 du\varphi^2(u, \omega) < \infty \qquad \text{a.s.} \ ,$$

the limit: $\displaystyle\lim_{\varepsilon \downarrow 0} \int_\varepsilon^1 \frac{du}{u} B_u \varphi(u, \omega)$ exists , since both limits, as $\varepsilon \to 0$, of

$\displaystyle\int_\varepsilon^1 dB_u \varphi(u, \omega)$ and $\displaystyle\int_\varepsilon^1 \beta_u \varphi(u, \omega)$ exist. This general existence result should be contrasted with the following

Lemma 1.2 Let $(\varphi(u, w); u \leq 1)$ be a $(\mathcal{G}_u)_{u \leq 1}$ predictable process such that: $\displaystyle\int_0^1 du\varphi^2(u, \omega) < \infty$ a.s. Then, the following properties are equivalent

(i) $\displaystyle\int_0^1 \frac{du}{\sqrt{u}} |\varphi(u, \omega)| < \infty;$ $\quad (ii)$ $\displaystyle\int_0^1 \frac{du}{u} |B_u|\, |\varphi(u, \omega)| < \infty;$

(iii) the process $\left(\displaystyle\int_0^t d\beta_u \varphi(u, \omega), t \leq 1 \right)$ is a $(\mathcal{B}_t, t \leq 1)$ semimartingale.

For a proof of this Lemma, we refer the reader to Jeulin-Yor ([53]); the equivalence between *(i)* and *(ii)* is a particular case of a useful lemma due to Jeulin ([51], p. 44).

(1.5.2) We now translate the above existence result, at least for $\varphi(u, \omega) = f(u)$, with f in $L^2([0, 1])$ in terms of a convergence result for certain integrals of the Ornstein-Uhlenbeck process.

Define the Ornstein-Uhlenbeck process with parameter $\mu \in \mathbb{R}$, as the unique solution of Langevin's equation:

$$X_t = x + B_t + \mu \int_0^t ds\, X_s \; ;$$

the method of variation of constants yields the formula:

$$X_t = e^{\mu t}\left(x + \int_0^t e^{-\mu s} dB_s\right) \; .$$

When $\mu = -\lambda$, with $\lambda > 0$, and x is replaced by a Gaussian centered variable X_0, with variance $\beta = \frac{1}{2\lambda}$, then the process:

$$Y_t = e^{-\lambda t}\left(X_0 + \int_0^t e^{\lambda s} dB_s\right)$$

is stationary, and may also be represented as:

$$Y_t = \frac{1}{\sqrt{2\lambda}} e^{-\lambda t} \tilde{B}_{e^{2\lambda t}} = \frac{1}{\sqrt{2\lambda}} e^{\lambda t} \hat{B}_{e^{-2\lambda t}} \; ,$$

where $(\tilde{B}_u)_{u \geq 0}$ and $(\hat{B}_u)_{u \geq 0}$ are two Brownian motions, which are linked by:

$$\tilde{B}_u = u \hat{B}_{1/u} \; .$$

We now have the following

Proposition 1.2 *For any $g \in L^2([0, \infty])$,* $\left(\int_0^t ds\, g(s) Y_s, t \to \infty\right)$ *converges a.s. and in L^2 (in fact, in every $L^p, p < \infty$).*

PROOF: Using the representation of $(Y_t, t \geq 0)$ in terms of $\hat{B}$, we have:

$$\int\limits_0^t ds \, g(s) Y_s = \int\limits_{e^{-2\lambda t}}^1 \frac{du}{u} \hat{B}_u \frac{1}{\sqrt{2\lambda u}} g\left(\frac{1}{2\lambda} \log \frac{1}{u}\right) \ .$$

Now, the application

$$g \longrightarrow \frac{1}{\sqrt{2\lambda u}} g\left(\frac{1}{2\lambda} \log \frac{1}{u}\right)$$
$$L^2([0, \infty]) \longrightarrow L^2([0, 1])$$

is an isomorphism of Hilbert spaces; the result follows. $\square$

1.6 Fourier transform and Brownian motion

There has been, since Lévy's discovery of local times, a lot of interest in the occupation measure of Brownian motion, that is, for fixed t and w, the measure $\lambda_{w,t}(dx)$ defined by:

$$\int \lambda_{w,t}(dx) f(x) = \int\limits_0^t ds \, f\left(B_s(w)\right) \ .$$

In particular, one may show that, a.s., the Fourier transform of $\lambda_{w,t}$, that is:
$\hat{\lambda}_{w,t}(\mu) \equiv \int\limits_0^t ds \exp(i\mu B_s(w))$ is in $L^2(d\mu)$; therefore, $\lambda_{w,t}(dx)$ is absolutely
continuous and its family of densities are the local times of B up to time t.
Now, we are interested in a variant of this, namely we consider:

$$\int\limits_0^t ds \, g(s) \exp(i\mu B_s), \quad \mu \neq 0 \ ,$$

where g satisfies: $\int\limits_0^t ds |g(s)| < \infty$, for every $t > 0$. We note the following

Proposition 1.3 *Let $\mu \in \mathbb{R}$, $\mu \neq 0$, and define: $\lambda = \frac{\mu^2}{2}$. Let $(Y_t, t \geq 0)$ be the stationary Ornstein-Uhlenbeck process, with parameter λ. Then, we have*

the following identities:

$$E\left[\left|\int_0^t ds\, g(s)\exp(i\mu B_s)\right|^2\right] = \mu^2 E\left[\left(\int_0^t ds\, g(s)Y_s\right)^2\right]$$

$$= \int_0^t ds \int_0^t du\, g(s)g(u)e^{-\lambda|u-s|} \ .$$

Corollary 1.3.1 *For any $\mu \neq 0$, and for any function $g \in L^2([0,\infty)]$,*
$$\left(\int_0^t ds\, g(s)\exp(i\mu B_s), t \to \infty\right) \quad \text{converges a.s. and in } L^2 \text{ (also in every}$$
$L^p, p < \infty).$

PROOF: The L^2 convergence follows immediately from the Proposition and from the L^2 convergence of the corresponding quantity for Y. The a.s. convergence is obtained from the martingale convergence theorem. Indeed, if we define: $\Gamma(\mu) = L^2\text{-}\lim\limits_{t\to\infty} \int_0^t ds\, g(s)e^{i\mu B_s}$, we have

$$E\left[\Gamma(\mu) \mid \mathcal{B}_t\right] = \int_0^t ds\, g(s)e^{i\mu B_s} + e^{i\mu B_t}\int_t^\infty ds\, g(s)e^{-\lambda(s-t)} \ .$$

The left-hand side converges a.s., hence, so does the right-hand side; but, the second term on the right-hand side goes to 0, since:

$$\left|e^{i\mu B_t}\int_t^\infty ds\, g(s)e^{-\lambda(s-t)}\right| \leq \left(\int_t^\infty ds\, g^2(s)\right)^{1/2}\frac{1}{\sqrt{2\lambda}} \xrightarrow[t\to\infty]{} 0 \ .$$

□

From the above results, the r.v. $\Gamma(\mu) \equiv \int_0^\infty ds\, g(s)\exp(i\mu B_s)$ is well-defined; it admits the following representation as a stochastic integral:

$$\Gamma(\mu) = \int_0^\infty ds\, g(s)\exp(-\lambda s) + i\mu \int_0^\infty dB_s \exp(i\mu B_s)G_\lambda(s) \ ,$$

where:

$$G_\lambda(s) = \int\limits_s^\infty du \, g(u) \exp -\lambda(u - s) \ .$$

Hence, $\Gamma(\mu)$ is the terminal variable of a martingale in the Brownian filtration, the increasing process of which is uniformly bounded. Therefore, we have:

$$E\left[\exp\left(\alpha|\Gamma(\mu)|^2\right)\right] < \infty, \quad \text{for } \alpha \text{ sufficiently small.}$$

Many properties of the variables $\Gamma(\mu)$ have been obtained by C. Donati-Martin [30].

Comments on Chapter 1

- In paragraph 1.1, some explicit and well-known realizations of the Brownian bridges are presented, with the help of the Gaussian character of Brownian motion.
- In paragraph 1.2, it is shown that the filtration of those Brownian bridges is that of a Brownian motion, and in paragraph 1.3, the application which transforms the original Brownian motion into the new one is shown to be ergodic; these two paragraphs follow Jeulin-Yor [54] closely.

One may appreciate how much the Gaussian structure facilitates the proofs in comparing the above development (Theorem 1.2, say) with the problem, not yet completely solved, of proving that Lévy's transformation:

$$(B_t, t \geq 0) \longrightarrow \left(\int\limits_0^t sgn(B_s)dB_s \ ; \ t \geq 0\right)$$

is ergodic. Dubins and Smorodinsky [35] have made some important progress on this question.
- Paragraph 1.4 is taken from Jeulin-Yor [54]; it is closely connected to works of H. Föllmer [43] and O. Brockhaus [22]. Also, the discussion and the results found in the same paragraph 1.4 look very similar to those in Carlen [23], but we have not been able to establish a precise connection between these two works.
- Paragraph 1.5 is taken mostly from Donati-Martin and Yor [32], whilst the content of paragraph 1.6 has been the starting point of Donati-Martin [30].

Chapter 2

The laws of some quadratic functionals of BM

In Chapter 1, we studied a number of properties of the Gaussian space of Brownian motion; this space may be seen as corresponding to the first level of complexity of variables which are measurable with respect to $\mathcal{F}_\infty \equiv \sigma\{B_s, s \geq 0\}$, where $(B_s, s \geq 0)$ denotes Brownian motion. Indeed, recall that N. Wiener proved that every $L^2(\mathcal{F}_\infty)$ variable X may be represented as:

$$X = E(X) + \sum_{n=1}^{\infty} \int_0^\infty dB_{t_1} \int_0^{t_1} dB_{t_2} \ldots \int_0^{t_{n-1}} dB_{t_n} \varphi_n(t_1, \ldots, t_n)$$

where φ_n is a deterministic Borel function which satisfies:

$$\int_0^\infty dt_1 \ldots \int_0^{t_{n-1}} dt_n \varphi_n^2(t_1, \ldots, t_n) < \infty .$$

In this Chapter, we shall study the laws of some of the variables X which correspond to the second level of complexity, that is: which satisfy $\varphi_n = 0$, for $n \geq 3$. In particular, we shall obtain the Laplace transforms of certain quadratic functionals of B, such as:

$$\alpha B_t^2 + \beta \int_0^t ds\, B_s^2 , \quad \int_0^t d\mu(s) B_s^2 , \quad \text{and so on...}$$

2.1 Lévy's area formula and some variants

(2.1.1) We consider $(B_t, t \geq 0)$ a δ-dimensional BM starting from $a \in \mathbb{R}^\delta$. We write $x = |a|^2$, and we look for an explicit expression of the quantity:

$$I_{\alpha,b} \overset{\text{def}}{=} E\left[\exp\left(-\alpha|B_t|^2 - \frac{b^2}{2}\int_0^t ds|B_s|^2\right)\right] .$$

We now show that, as a consequence of Girsanov's transformation, we may obtain the following formula[1] for $I_{\alpha,b}$:

$$I_{\alpha,b} = \left(\text{ch}(bt) + 2\frac{\alpha}{b}\text{sh}(bt)\right)^{-\delta/2} \exp{-\frac{xb\left(1 + \frac{2\alpha}{b}\coth bt\right)}{2\left(\coth(bt) + \frac{2\alpha}{b}\right)}} \qquad (2.1)$$

PROOF: We may assume that $b \geq 0$. We consider the new probability $P^{(b)}$ defined by:

$$P^{(b)}_{|\mathcal{F}_t} = \exp\left\{-\frac{b}{2}\left(|B_t|^2 - x - \delta t\right) - \frac{b^2}{2}\int_0^t ds|B_s|^2\right\} \cdot P_{|\mathcal{F}_t} .$$

Then, under $P^{(b)}$, $(B_u, u \leq t)$ satisfies the following equation,

$$B_u = a + \beta_u - b\int_0^u ds\, B_s , \qquad u \leq t ,$$

where $(\beta_u, u \leq t)$ is a $(P^{(b)}, \mathcal{F}_t)$ Brownian motion.

Hence, $(B_u, u \leq t)$ is an Ornstein-Uhlenbeck process with parameter $-b$, starting from a. Consequently, $(B_u, u \leq t)$ may be expressed explicitly in terms of β, as

$$B_u = e^{-bu}\left(a + \int_0^u e^{bs}d\beta_s\right) , \qquad (2.2)$$

a formula from which we can immediately compute the mean and the variance of the Gaussian variable B_u (considered under $P^{(b)}$). This clearly solves the problem, since we have:

---

[*] Throughout the volume, we use the French abbreviations ch, sh, th for, respectively, cosh, sinh, tanh,...

$$I_{\alpha,b} = E^{(b)}\left[\exp\left(-\alpha|B_t|^2 + \frac{b}{2}\left(|B_t|^2 - x - \delta t\right)\right)\right] \ ,$$

and formula (2.1) now follows from some straightforward, if tedious, computations. □

Exercise 2.1: Show that $\exp\left\{\dfrac{b}{2}\left(|B_t|^2 - x - \delta t\right) - \dfrac{b^2}{2}\displaystyle\int_0^t ds|B_s|^2\right\}$ is also a $(P, (\mathcal{F}_t))$ martingale, and that we might have considered this martingale as a Radon-Nikodym density to arrive to the same formula (2.1).

(2.1.2) The same method allows to compute the joint Fourier-Laplace transform of the pair: $\left(\displaystyle\int_0^t f(u)dB_u, \ \displaystyle\int_0^t du\, B_u^2\right)$ where for simplicity, we take here the dimension δ to be 1.

Indeed, to compute:

$$E\left[\exp\left(i\int_0^t f(u)dB_u - \frac{b^2}{2}\int_0^t du\, B_u^2\right)\right] \ , \tag{2.3}$$

all we need to know, via the above method, is the joint distribution of $\displaystyle\int_0^t f(u)dB_u$ and B_t, under $P^{(b)}$.

This is clearly equivalent to being able to compute the mean and variance of $\displaystyle\int_0^t g(u)dB_u$, for any $g \in L^2([0,t], du)$.

However, thanks to the representation (2.2), we have:

$$\int_0^t g(u)dB_u = \int_0^t g(u)\left\{-be^{-bu}du \cdot \left(a + \int_0^u e^{bs}d\beta_s\right) + e^{-bu}(e^{bu}d\beta_u)\right\}$$

$$= -ba\int_0^t g(u)e^{-bu}du + \int_0^t d\beta_u\left(g(u) - e^{bu}b\int_u^t e^{-bs}g(s)ds\right) \ .$$

Hence, the mean of $\displaystyle\int_0^t g(u)dB_u$ under $P^{(b)}$ is: $\displaystyle -ba\int_0^t g(u)e^{-bu}du$, and its

variance is: $\displaystyle \int_0^t du \left(g(u) - be^{bu}\int_u^t e^{-bs}g(s)ds \right)^2.$

We shall not continue the discussion at this level of generality, but instead, we indicate one example where the computations have been completely carried out.

The next formulae will be simpler if we work in a two-dimensional setting; therefore, we shall consider $Z_u = X_u + iY_u$, $u \geq 0$, a $\mathbb{C}$-valued BM starting from 0, and we define $G = \displaystyle\int_0^1 ds\, Z_s$, the barycenter of Z over the time-interval [0,1].

The above calculations lead to the following formula (taken with small enough $\rho, \sigma \geq 0$):

$$E\left[\exp -\frac{\lambda^2}{2}\left(\int_0^1 ds|Z_s|^2 - \rho|G|^2 - \sigma|Z_1|^2 \right) \right]$$
$$= \left\{ (1-\rho)\text{ch}\lambda + \rho\frac{\text{sh}\lambda}{\lambda} + \sigma\left[(\rho-1)\lambda\text{sh}\lambda - 2\rho(\text{ch}\lambda - 1) \right] \right\}^{-1} \qquad (2.4)$$

which had been obtained by a different method by Chan-Dean-Jansons-Rogers [26].

(2.1.3)   Before we continue with some consequences of formulae (2.1) and (2.4), let us make some remarks about the above method:

it consists in changing probability so that the quadratic functional disappears, and the remaining problem is to compute the mean and variance of a Gaussian variable. Therefore, this method consists in transfering some computational problem for a variable belonging to (the first and) the second Wiener chaos to computations for a variable in the first chaos; in other words, it consists in a *linearization* of the original problem.

In the last paragraph of this Chapter, we shall use this method again to deal with the more general problem, when $\displaystyle\int_0^t ds|B_s|^2$ is replaced by $\displaystyle\int_0^t d\mu(s)|B_s|^2$.

(2.1.4) A number of computations found in the literature can be obtained very easily from the formulae (2.1) and (2.4).

a) The following formula is easily deduced from formula (2.1):

$$
E_a \left[\exp - \frac{b^2}{2} \int_0^t ds |B_s|^2 \mid B_t = 0 \right] = E_0 \left[\exp - \frac{b^2}{2} \int_0^t ds |B_s|^2 \mid B_t = a \right]
$$

$$
= \left(\frac{bt}{\operatorname{sh}(bt)} \right)^{\delta/2} \exp - \frac{|a|^2}{2t} (bt \coth(bt) - 1)
$$

(2.5)

which, in the particular case $a = 0$, yields the formula:

$$
E_0 \left[\exp \left(- \frac{b^2}{2} \int_0^t ds |B_s|^2 \right) \mid B_t = 0 \right] = \left(\frac{bt}{\operatorname{sh}(bt)} \right)^{\delta/2}
$$

(2.6)

Lévy's formula for the stochastic area

$$
A_t \stackrel{\text{def}}{=} \int_0^t (X_s dY_s - Y_s dX_s)
$$

of planar Brownian motion $B_t = (X_t, Y_t)$ may now be deduced from formula (2.5); precisely, one has:

$$
E_0 \left[\exp(ibA_t) \mid B_t = a \right] = \left(\frac{bt}{\operatorname{sh} bt} \right) \exp - \frac{|a|^2}{2t} (bt \coth bt - 1)
$$

(2.7)

To prove formula (2.7), first remark that, thanks to the rotational invariance of the law of Brownian motion (starting from 0), we have:

$$
E_0 \left[\exp(ibA_t) \mid B_t = a \right] = E_0 \left[\exp(ibA_t) \mid |B_t| = |a| \right] ,
$$

and then, we can write:

$$
A_t = \int_0^t |B_s| d\gamma_s ,
$$

where $(\gamma_t, t \geq 0)$ is a one dimensional Brownian motion independent from $(|B_s|, s \geq 0)$. Therefore, we obtain:

$$E_0\left[\exp(ib\mathcal{A}_t) \mid |B_t| = |a|\right] = E_0\left[\exp\left(-\frac{b^2}{2}\int_0^t ds|B_s|^2\right) \mid B_t = a\right]$$

and formula (2.7) is now deduced from formula (2.5).

b) Similarly, from formula (2.4), one deduces:

$$E\left[\exp\left(-\frac{\mu^2}{2}\int_0^1 ds|Z_s - G|^2\right) \mid Z_1 = z\right]$$

$$= \left(\frac{\mu/2}{\mathrm{sh}\mu/2}\right)^2 \exp-\frac{|z|^2}{2}\left(\frac{\mu}{2}\coth\frac{\mu}{2} - 1\right) \qquad (2.8)$$

c) As yet another example of application of the method, we now derive the following formula obtained by M. Wenocur [91] (see also, in the same vein, [92]):

consider $(W(t), t \geq 0)$ a 1-dimensional BM, starting from 0, and define: $X_t = W_t + \mu t + x$, so that $(X_t, t \geq 0)$ is the Brownian motion with drift μ, starting from x.

Then, M. Wenocur [91] obtained the following formula:

$$E\left[\exp\left(-\frac{\lambda^2}{2}\int_0^1 ds\, X_s^2\right)\right] = \frac{1}{(\mathrm{ch}\lambda)^{1/2}}\exp(H(x, \mu, \lambda)) , \qquad (2.9)$$

where

$$H(x, \mu, \lambda) = -\frac{\mu^2}{2}\left(1 - \frac{\mathrm{th}\lambda}{\lambda}\right) - x\mu\left(1 - \frac{1}{\mathrm{ch}\lambda}\right) - \frac{x^2}{2}\lambda\mathrm{th}\lambda .$$

We shall now sketch a proof of this formula, by applying twice Girsanov's theorem. First of all, we may "get rid of the drift μ", since:

$$E\left[\exp\left(-\frac{\lambda^2}{2}\int_0^1 ds\, X_s^2\right)\right]$$

$$= E_x\left[\exp\left(\mu(X_1 - x) - \frac{\mu^2}{2}\right)\exp-\frac{\lambda^2}{2}\int_0^1 ds\, X_s^2\right]$$

where P_x denotes the law of Brownian motion starting from x. We apply Girsanov's theorem a second time, thereby replacing P_x by $P_x^{(\lambda)}$, the law of the Ornstein-Uhlenbeck process, with parameter λ, starting from x. We then obtain:

$$
E_x \left[\exp \left(\mu X_1 - \frac{\lambda^2}{2} \int_0^1 ds\, X_s^2 \right) \right]
$$

$$
= E_x^{(\lambda)} \left[\exp \left(\mu X_1 + \frac{\lambda}{2} X_1^2 \right) \exp \left(-\frac{\lambda}{2} (x^2 + 1) \right) \right] \;,
$$

and it is now easy to finish the proof of (2.9), since, as shown at the beginning of this paragraph, the mean and variance of X_1 under $P_x^{(\lambda)}$ are known.

Exercise 2.2: 1) Extend formula (2.9) to a δ-dimensional Brownian motion with constant drift.

2) Derive formula (2.1) from this extended formula (2.9).

Hint: Integrate both sides of the extended formula (2.9) with respect to $d\mu \exp - \left(c|\mu|^2 \right)$ on $\mathbb{R}^\delta$.

Exercise 2.3: Let $(B_t, t \geq 0)$ be a 3-dimensional Brownian motion starting from 0.

1. Prove the following formula:
   for every $m \in \mathbb{R}^3$, $\xi \in \mathbb{R}^3$ with $|\xi| = 1$, and $\lambda \in \mathbb{R}^*$,

$$
E \left[\exp \left(i\lambda \xi \cdot \int_0^1 B_s \times dB_s \right) \Big| B_1 = m \right]
$$

$$
= \left(\frac{\lambda}{\mathrm{sh}\lambda} \right) \exp \left(\frac{|m|^2 - (\xi \cdot m)^2}{2} (1 - \lambda \coth \lambda) \right) \;,
$$

where $x \cdot y$, resp.: $x \times y$, denotes the scalar product, resp.: the vector product, of x and y in $\mathbb{R}^3$.

Hint: Express $\xi \cdot \int_0^1 B_s \times dB_s$ in terms of the stochastic area of the 2-dimensional Brownian motion: $(\eta \cdot B_s; (\xi \times \eta) \cdot B_s; s \geq 0)$ where η is a suitably chosen unit vector of $\mathbb{R}^3$, which is orthogonal to ξ.

2. Prove that, for any $\lambda \in \mathbb{R}^*$, $z \in \mathbb{R}^3$, and $\xi \in \mathbb{R}^3$, with $|\xi| = 1$, one has:

$$E\left[\exp i\left(z \cdot B_1 + \lambda\xi \cdot \int_0^1 B_s \times dB_s\right)\right]$$
$$= \frac{1}{(\mathrm{ch}\lambda)} \exp -\frac{1}{2}\left(|z|^2\frac{\mathrm{th}\lambda}{\lambda} + (z \cdot \xi)^2\left(1 - \frac{\mathrm{th}\lambda}{\lambda}\right)\right) .$$

2.2 Some identities in law and an explanation of them via Fubini's theorem

(2.2.1)　We consider again formula (2.4), in which we take $\rho = 1$, and $\sigma = 0$. We then obtain:

$$E\left[\exp\left(-\frac{\lambda^2}{2}\int_0^1 ds|Z_s - G|^2\right)\right] = \frac{\lambda}{\mathrm{sh}\lambda} ,$$

but, from formula (2.6), we also know that, using the notation $(\tilde{Z}_s, s \leq 1)$ for the complex Brownian bridge of length 1:

$$E\left[\exp\left(-\frac{\lambda^2}{2}\int_0^1 ds|\tilde{Z}_s|^2\right)\right] = \frac{\lambda}{\mathrm{sh}\lambda} ;$$

hence, the following identity in law holds:

$$\int_0^1 ds|Z_s - G|^2 \overset{\text{(law)}}{=} \int_0^1 ds|\tilde{Z}_s|^2 , \tag{2.10}$$

an identity which had been previously noticed by several authors (see, e.g., [33]).

Obviously, the fact that, in (2.10), Z, resp. $\tilde{Z}$, denotes a complex valued BM, resp. Brownian bridge, instead of a real-valued process, is of no importance, and (2.10) is indeed equivalent to:

$$\int_0^1 dt(B_t - G)^2 \overset{\text{(law)}}{=} \int_0^1 dt\tilde{B}_t^2 , \tag{2.11}$$

where $(B_t, t \leq 1)$, resp. $(\tilde{B}_t, t \leq 1)$ now denotes a 1-dimensional BM, resp. Brownian bridge, starting from 0.

(2.2.2) Our first aim in this paragraph is to give a simple explanation of (2.11) via Fubini's theorem.

Indeed, if B and C denote two independent Brownian motions and $\varphi \in L^2([0,1], du\, ds)$, we have:

$$\int_0^1 dB_u \int_0^1 dC_s \varphi(u,s) \stackrel{\text{a.s.}}{=} \int_0^1 dC_s \int_0^1 dB_u \varphi(u,s) \ ,$$

which, as a corollary, yields:

$$\int_0^1 du \left(\int_0^1 dC_s \varphi(u,s) \right)^2 \stackrel{\text{(law)}}{=} \int_0^1 du \left(\int_0^1 dC_s \varphi(s,u) \right)^2 \qquad (2.12)$$

(in the sequel, we shall refer to this identity as to the "Fubini-Wiener identity in law").

The identity (2.11) is now a particular instance of (2.12), as the following Proposition shows.

Proposition 2.1 *Let* $f : [0,1] \to \mathbb{R}$ *be a* C^1-*function such that* $f(1) = 1$. *Then, we have:*

$$\int_0^1 ds \left(B_s - \int_0^1 dt\, f'(t) B_t \right)^2 \stackrel{\text{(law)}}{=} \int_0^1 ds (B_s - f(s)B_1)^2 \ . \qquad (2.13)$$

In particular, in the case $f(s) = s$, *we obviously recover (2.11).*

PROOF: It follows from the identity in law (2.12), where we take:

$$\varphi(s,u) = \big(1_{(u \leq s)} - (f(1) - f(u))\big) 1_{((s,u) \in [0,1]^2)} \ . \qquad \square$$

Here is another variant, due to Shi Zhan, of the identity in law (2.13).

Exercise 2.4: Let $\mu(dt)$ be a probability on $\mathbb{R}_+$. Then, prove that:

$$\int\limits_0^\infty \mu(dt) \left(B_t - \int\limits_0^\infty \mu(ds)B_s \right)^2 \overset{\text{(law)}}{=} \int\limits_0^\infty \tilde{B}^2_{\mu[0,t]}dt \ ,$$

where $(\tilde{B}_u, u \leq 1)$ is a standard Brownian bridge.

As a second application of (2.12), or rather of a discrete version of (2.12), we prove a striking identity in law (2.14), which resembles the integration by parts formula.

Theorem 2.1 *Let $(B_t, t \geq 0)$ be a 1-dimensional BM starting from 0. Let $0 \leq a \leq b < \infty$, and $f, g : [a, b] \to \mathbb{R}_+$ be two continuous functions, with f decreasing, and g increasing.*

$$\int\limits_a^b -df(x)B^2_{g(x)} + f(b)B^2_{g(b)} \overset{\text{(law)}}{=} g(a)B^2_{f(a)} + \int\limits_a^b dg(x)B^2_{f(x)} \ . \qquad (2.14)$$

In order to prove (2.14), it suffices to show that the identity in law:

$$-\sum_{i=1}^n (f(t_{i+1}) - f(t_i))B^2_{g(t_i)} + f(t_n)B^2_{g(t_n)}$$

$$\overset{\text{(law)}}{=} g(t_1)B^2_{f(t_1)} + \sum_{i=2}^n (g(t_i) - g(t_{i-1}))B^2_{f(t_i)} \ , \qquad (2.15)$$

where $a = t_1 < t_2 < \cdots < t_n = b$, holds, and then to let the mesh of the subdivision tend to 0. .

Now, (2.15) is a particular case of a discrete version of (2.12), which we now state.

Theorem 2.2 *Let $\mathbf{X}_n = (X_1, \ldots, X_n)$ be an n-dimensional Gaussian vector, the components of which are independent, centered, with variance 1. Then, for any $n \times n$ matrix A, we have:*

$$|A\mathbf{X}_n| \overset{\text{(law)}}{=} |A^*\mathbf{X}_n| \ ,$$

where A^* is the transpose of A, and, if $\mathbf{x}_n = (x_1, \text{---}, x_n) \in \mathbb{R}^n$, we denote:
$$|\mathbf{x}_n| = \left(\sum_{i=1}^n x_i^2 \right)^{1/2} \ .$$

Corollary 2.2.1 *Let* $(Y_1, \ldots, Y_n)$ *and* $(Z_1, \ldots, Z_n)$ *be two* n*-dimensional Gaussian vectors such that*

i) $Y_1, Y_2 - Y_1, \ldots, Y_n - Y_{n-1}$ *are independent;*

ii) $Z_n, Z_n - Z_{n-1}, \ldots, Z_2 - Z_1$ *are independent.*

Then, we have

$$-\sum_{i=1}^{n} Y_i^2 \left(E(Z_{i+1}^2) - E(Z_i^2) \right) \stackrel{\text{(law)}}{=} \sum_{i=1}^{n} Z_i^2 \left(E(Y_i^2) - E(Y_{i-1}^2) \right) \qquad (*)$$

where we have used the convention: $E(Z_{n+1}^2) = E(Y_0^2) = 0$.

The identity in law (2.15) now follows as a particular case of $(*)$.

2.3 The laws of squares of Bessel processes

Consider $(B_t, t \geq 0)$ a δ-dimensional ($\delta \in \mathbb{N}$, for the moment...) Brownian motion starting from a, and define: $X_t = |B_t|^2$. Then, $(X_t, t \geq 0)$ satisfies the following equation

$$X_t = x + 2 \int_0^t \sqrt{X_s} d\beta_s + \delta t \ , \qquad (2.16)$$

where $x = |a|^2$, and $(\beta_t, t \geq 0)$ is a 1-dimensional Brownian motion. More generally, from the theory of 1-dimensional stochastic differential equations, we know that for any pair $x, \delta \geq 0$, the equation (2.16) admits one strong solution, hence, a fortiori, it enjoys the uniqueness in law property.

Therefore, we may define, on the canonical space $\Omega_+^* \equiv C(\mathbb{R}_+, \mathbb{R}_+)$, Q_x^δ as the law of a process which satisfies (2.16).

The family $(Q_x^\delta, x \geq 0, \delta \geq 0)$ possesses the following additivity property, which is obvious for integer dimensions.

Theorem 2.3 *(Shiga-Watanabe [83]) For any $\delta, \delta', x, x' \geq 0$, the identity:*

$$Q_x^\delta * Q_{x'}^{\delta'} = Q_{x+x'}^{\delta+\delta'}$$

holds, where $*$ denotes the convolution of two probabilities on Ω_+^*.

Now, for any positive, σ-finite, measure μ on $\mathbb{R}_+$, we define:

$$I_\mu(\omega) = \int\limits_0^\infty d\mu(s) X_s(\omega) \ ,$$

and we deduce from the theorem that there exist two positive constants $A(\mu)$ and $B(\mu)$ such that:

$$Q_x^\delta \left(\exp -\frac{1}{2} I_\mu \right) = (A(\mu))^x (B(\mu))^\delta \ .$$

The next theorem allows to compute $A(\mu)$ and $B(\mu)$.

Theorem 2.4 *For any ≥ 0 Radon measure μ on $[0, \infty)$, one has:*

$$Q_x^\delta \left(\exp -\frac{1}{2} I_\mu \right) = (\phi_\mu(\infty))^{\delta/2} \exp \left(\frac{x}{2} \phi_\mu^+(0) \right) \ ,$$

where ϕ_μ denotes the unique solution of:

$$\phi'' = \mu\phi \ \text{ on } \ (0, \infty) \ , \quad \phi_\mu(0) = 1, 0 \leq \phi \leq 1 \ ,$$

and $\phi_\mu^+(0)$ is the right derivative of ϕ_μ at 0.

PROOF: For simplicity, we assume that μ is diffuse, and that its support is contained in $(0, 1)$.

Define: $F_\mu(t) = \dfrac{\phi_\mu'(t)}{\phi_\mu(t)}$, and $\hat{F}_\mu(t) = \int\limits_0^t \dfrac{\phi_\mu'(s) ds}{\phi_\mu(s)} = \log \phi_\mu(t)$.

Then, remark that:

$$Z_t^\mu \stackrel{\text{def}}{=} \exp \left\{ \frac{1}{2} \left[F_\mu(t) X_t - F_\mu(0) x - \delta \hat{F}_\mu(t) \right] - \frac{1}{2} \int\limits_0^t X_s d\mu(s) \right\}$$

is a Q_x^δ-martingale, since it may be written as:

$$\exp\left\{\int_0^t F_\mu(s)dM_s - \frac{1}{2}\int_0^t F_\mu^2(s)d\langle M\rangle_s\right\} ,$$

where: $M_t = \frac{1}{2}(X_t - \delta t)$, and $\langle M\rangle_t = \int_0^t ds\, X_s.$

It now remains to write: $Q_x^\delta(Z_1^\mu) = 1$, and to use the fact that $F_\mu(1) = 0$ to obtain the result stated in the theorem. □

Exercise 2.5: 1) Prove that the integration by parts formula (2.14) can be extended as follows:

$$(*) \quad \int_a^b -df(x)X_{g(x)} + f(b)X_{g(b)} \overset{\text{(law)}}{=} g(a)X_{f(a)} + \int_a^b dg(x)X_{f(x)} ,$$

where X is a BESQ process, with any strictly positive dimension, starting from 0.

2) Prove the following convergence in law result:

$$\left(\sqrt{n}\left(\frac{1}{n}X_t^{(n)} - t\right), t \geq 0\right) \xrightarrow[n\to\infty]{\text{(law)}} (c\beta_{t^2}; t \geq 0) ,$$

for a certain constant $c > 0$, where $(X_t^{(n)}, t \geq 0)$ denotes a BESQ^n process, starting from 0, and $(\beta_t, t \geq 0)$ denotes a real-valued BM, starting from 0.

3) Prove that the process $(X_t \equiv \beta_{t^2}, t \geq 0)$ satisfies $(*)$.

Comments on Chapter 2

For many reasons, a number of computations of the Laplace or Fourier transform of the distribution of quadratic functionals of Brownian motion, or related processes, are being published almost every year; the origins of the interests in such functionals range from Bismut's proof of the Atiyah-Singer theorem, to polymer studies (see Chan-Dean-Jansons-Rogers [26] for the latter).

Duplantier [36] presents a good list of references to the literature.

The methods used by the authors to obtain closed formulae for the corresponding characteristic functions or Laplace transforms fall essentially into one of the three following categories:

i) P. Lévy's *diagonalisation procedure,* which has a strong functional analysis flavor; this method may be applied very generally and is quite powerful; however, the characteristic functions or Laplace transforms then appear as infinite products, which have to be recognized in terms of, say, hyperbolic functions...

ii) the *change of probability method* which, in effect, linearizes the problem, i.e.: it allows to transform the study of a quadratic functional into the computation of the mean and variance of an adequate Gaussian variable; paragraph 2.1 above gives an important example of this method.

iii)finally, the *reduction method,* which simply consists in trying to reduce the computation for a certain quadratic functional to similar computations which have already been done. Exercise 2.3, and indeed the whole paragraph 2.2 above give some examples of application. The last formula in Exercise 2.3 is due to Foschini and Shepp [44] and the whole exercise is closely related to the work of Berthuet [6] on the stochastic volume of $(B_u, u \leq 1)$.

Paragraph 2.3 is closely related to Pitman-Yor ([73], [74]).

Some extensions of the integration by parts formula (2.14) to stable processes and some converse studies have been made by Donati-Martin, Song and Yor [31].

Chapter 3

Squares of Bessel processes and Ray-Knight theorems for Brownian local times

Chapters 1 and 2 were devoted to the study of some properties of variables in the first and second Wiener chaos. In the present Chapter, we are studying variables which are definitely at a much higher level of complexity in the Wiener chaos decomposition; in fact, they have infinitely many Wiener chaos components.

More precisely, we shall study, in this Chapter, some properties of the Brownian local times, which may be defined by the *occupation times formula:*

$$\int\limits_0^t ds\, f(B_s) = \int\limits_{-\infty}^\infty da\, f(a)\ell_t^a \ , \quad f \in b\left(\mathcal{B}(\mathbb{R})\right) \ ,$$

and, from Trotter's theorem, we may, and we shall, choose the family $(\ell_t^a; a \in \mathbb{R}, t \geq 0)$ to be jointly continuous.

This occupation times formula transforms an *integration in time,* into an *integration in space,* and it may be asked: what becomes of the Markov property through this change from time to space?

In fact, the Ray-Knight theorems presented below show precisely that there is some Markov property in space, that is: at least for some suitably chosen stopping times T, the process $(\ell_T^a, a \in \mathbb{R})$ is a strong Markov process, the law of which can be described precisely.

More generally, we shall try to show some evidence, throughout this Chapter, of a general *transfer principle* from time to space, which, in our opinion, permeates the various developments made around the Ray-Knight theorems on Brownian local times.

3.1 The basic Ray-Knight theorems

There are two such theorems, the first one being related to $T \equiv \tau_x$ $= \inf\{t \geq 0 : \ell_t^0 = x\}$, and the second one to $T' \equiv T_1 = \inf\{t : B_t = 1\}$.

(RK1) *The processes* $(\ell_{\tau_x}^a; a \geq 0)$ *and* $(\ell_{\tau_x}^{-a}; a \geq 0)$ *are two independent squares, starting at* x, *of 0-dimensional Bessel processes, i.e.: their common law is* Q_x^0.

(RK2) *The process* $(\ell_{T_1}^{1-a}; 0 \leq a \leq 1)$ *is the square of a 2-dimensional Bessel process starting from 0, i.e.: its law is* Q_0^2.

There are several important variants of (RK2), among which the two following ones.

(RK2)(a) *If* $(R_3(t), t \geq 0)$ *denotes the 3-dimensional Bessel process starting from 0, then the law of* $(\ell_\infty^a(R_3), a \geq 0)$ *is* Q_0^2.

(RK2)(b) *The law of* $(\ell_\infty^a(|B| + \ell^0); a \geq 0)$ *is* Q_0^2.

We recall that (RK2)(a) follows from (RK2), thanks to Williams' time reversal result:
$$(B_t; t \leq T_1) \stackrel{(\text{law})}{=} (1 - R_3(L_1 - t); t \leq L_1) \ ,$$
where $L_1 = \sup\{t > 0 : R_3(t) = 1\}$.

Then, (RK2)(b) follows from (RK2)(a) thanks to Pitman's representation of R_3 (see [71]), which may be stated as

$$(R_3(t), t \geq 0) \stackrel{(\text{law})}{=} (|B_t| + \ell_t^0; t \geq 0)$$

We now give a first example of the *transfer principle* from time to space mentioned above. Consider, for $\mu \in \mathbb{R}$, the solution of:

$$(*) \qquad\qquad X_t = B_t + \mu \int_0^t ds\, 1_{(X_s > 0)} \ ,$$

and call $P^{\mu,+}$ the law of this process on the canonical space Ω^*; in the following, we simply write P for the standard Wiener measure.

Then, from Girsanov's theorem, we have:

$$P^{\mu,+}_{|\mathcal{F}_t} = \exp\left\{ \mu \int_0^t 1_{(X_s>0)} dX_s - \frac{\mu^2}{2} \int_0^t ds\, 1_{(X_s>0)} \right\} \cdot P_{|\mathcal{F}_t}$$

$$= \exp\left\{ \mu \left(X_t^+ - \frac{1}{2}\ell_t^0 \right) - \frac{\mu^2}{2} \int_0^t ds\, 1_{(X_s>0)} \right\} \cdot P_{|\mathcal{F}_t} \quad,$$

where $(X_t)_{t\geq 0}$ denotes the canonical process on Ω^*, and $(\ell_t^0)_{t\geq 0}$ its local time at 0 (which is well defined P a.s.).

It follows from the above Radon-Nikodym relationship that, for any ≥ 0 measurable functional F on Ω_+^*, we have:

$$E^{\mu,+}\left[F(\ell_{T_1}^{1-a}; 0 \leq a \leq 1) \right]$$

$$= E\left[F(\ell_{T_1}^{1-a}; 0 \leq a \leq 1) \exp\left\{ -\frac{\mu}{2}(\ell_{T_1}^0 - 2) - \frac{\mu^2}{2} \int_0^1 da\, \ell_{T_1}^a \right\} \right]$$

$$(\dagger) \quad = Q_0^2\left[F(Z_a; 0 \leq a \leq 1) \exp\left\{ -\frac{\mu}{2}(Z_1 - 2) - \frac{\mu^2}{2} \int_0^1 da\, Z_a \right\} \right]$$

where $(Z_a, a \geq 0)$ now denotes the canonical process on Ω_+^* (to avoid confusion with X on Ω^*). The last equality follows immediately from (RK2).

Now, the exponential which appears as a Radon-Nikodym density in ($\dagger$) transforms Q_0^2 into $^{(-\mu)}Q_0^2$, a probability which is defined in the statement of Theorem 3.1 below (see paragraph 6 of Pitman-Yor [73] for details).

Hence, we have just proved the following

Theorem 3.1 *If $X^{(\mu)}$ denotes the solution of the equation (∗) above, then, the law of $\left(\ell_{T_1}^{1-a}(X^{(\mu)}); 0 \leq a \leq 1\right)$ is $^{(-\mu)}Q_0^2$, where $^\beta Q_x^\delta$ denotes the law of the square, starting at x, of the norm of a δ-dimensional Ornstein-Uhlenbeck process with parameter β, i.e.: a diffusion on $\mathbb{R}_+$ whose infinitesimal generator is:*

$$2y\frac{d^2}{dy^2} + (2\beta y + \delta)\frac{d}{dy}.$$

3.2 The Lévy-Khintchine representation of Q_x^δ

We have seen, in the previous Chapter, that for any $x, \delta \geq 0$, Q_x^δ is infinitely divisible (Theorems 2.3 and 2.4). We are now able to express its Lévy-Khintchine representation as follows

Theorem 3.2 *For any Borel function $f : \mathbb{R}_+ \to \mathbb{R}_+$, and $\omega \in \Omega_+^*$, we set*

$$I_f(\omega) = \langle \omega, f \rangle = \int_0^\infty dt\, \omega(t) f(t) \quad \text{and} \quad f_u(t) = f(u+t) \ .$$

Then, we have, for every $x, \delta \geq 0$:

$$Q_x^\delta(\exp - I_f)$$

$$= \exp - \int \mathbf{M}(d\omega) \left\{ x\left[1 - \exp(-I_f(\omega))\right] + \delta \int_0^\infty du(1 - \exp - I_{f_u}(\omega)) \right\} \ ,$$

where $\mathbf{M}(d\omega)$ is the image of the Itô measure $\mathbf{n}_+$ of positive excursions by the application which associates to an excursion ε the process of its local times:

$$\varepsilon \to (\ell_R^x(\varepsilon); x \geq 0) \ .$$

Before we give the proof of the theorem, we make some comments about the representations of Q_x^0 and Q_0^δ separately:
obviously, from the theorem, the representing measure of Q_x^0 is $x\mathbf{M}(d\omega)$, whereas the representing measure of Q_0^δ is $\delta\mathbf{N}(d\omega)$, where $\mathbf{N}(d\omega)$ is characterized by:

$$\int \mathbf{N}(d\omega)\left(1 - e^{-I_f(\omega)}\right) = \int \mathbf{M}(d\omega) \int_0^\infty du \left(1 - e^{-I_{f_u}(\omega)}\right)$$

and it is not difficult to see that this formula is equivalent to:

$$\int \mathbf{N}(d\omega)F(\omega) = \int \mathbf{M}(d\omega) \int_0^\infty du\, F\left(\omega((\cdot - u)^+)\right) \ ,$$

for any measurable ≥ 0 functional F.

Now, in order to prove the theorem, all we need to do is to represent Q_x^0, and Q_0^δ, for some dimension δ; in fact, we shall use (RK1) to represent Q_x^0, and (RK2) (b) to represent Q_0^2.

Our main tool will be (as is to be expected!) excursion theory. We first state the following consequences of the *master formulae* of excursion theory (see [81], Chapter XII, Propositions (1.10) and (1.12)).

Proposition 3.1 *Let $(M_t, t \geq 0)$ be a bounded, continuous process with bounded variation on compacts of $\mathbb{R}_+$, such that: $1_{(B_t=0)}dM_t = 0$.*

Then, (i) *if, moreover, $(M_t, t \geq 0)$ is a multiplicative functional, we have:*

$$E[M_{\tau_x}] = \exp\left(-x \int \mathbf{n}(d\varepsilon)(1 - M_R(\varepsilon))\right) ,$$

where $\mathbf{n}(d\varepsilon)$ denotes the Itô characteristic measure of excursions.

(ii) *More generally, if the multiplicativity property assumption is replaced by: $(M_t, t \geq 0)$ is a skew multiplicative functional, in the following sense:*

$$M_{\tau_s} = M_{\tau_{s-}}(M_R^{(s)}) \circ \theta_{\tau_{s-}} \qquad (s \geq 0) ,$$

for some measurable family of r.v.'s $(M_R^{(s)}; s \geq 0)$, then the previous formula should be modified as

$$E[M_{\tau_x}] = \exp\left(-\int\limits_0^x ds \int \mathbf{n}(d\varepsilon)\left(1 - M_R^{(s)}(\varepsilon)\right)\right) .$$

Taking $M_t \equiv \exp - \int\limits_0^t ds\, f(B_s, \ell_s)$, for $f : \mathbb{R} \times \mathbb{R}_+ \to \mathbb{R}$, a Borel function, we obtain, as an immediate consequence of the Proposition, the following important formula:

$$(*) \qquad E\left[\exp\left(-\int_0^{\tau_x} ds\, f(B_s, \ell_s)\right)\right]$$

$$= \exp-\int_0^x ds \int \mathbf{n}(d\varepsilon)\left(1 - \exp-\int_0^R du\, f(\varepsilon(u), s)\right).$$

As an application, if we take $f(y, \ell) \equiv 1_{(y\geq 0)}g(y)$, then the left-hand side of $(*)$ becomes:

$$Q_x^0(\exp -I_g)\ , \qquad \text{thanks to (RK1)},$$

while the right-hand side of $(*)$ becomes:

$$\exp -x\int \mathbf{n}_+(d\varepsilon)\left(1 - \exp-\int_0^R du\, g(\varepsilon(u))\right) = \exp -x\int \mathbf{M}(d\omega)\left(1 - e^{-I_g(\omega)}\right)$$

from the definition of $\mathbf{M}$.

Next, if we write formula $(*)$ with $f(y, \ell) = g(|y| + \ell)$, and $x = \infty$, the left-hand side becomes:

$$Q_0^2(\exp -I_g)\ , \qquad \text{thanks to (RK2) (b)},$$

while the right-hand side becomes:

$$\exp -\int_0^\infty ds \int \mathbf{n}(d\varepsilon)\left(1 - \exp-\int_0^R du\, g(|\varepsilon(u)| + s)\right)$$

$$= \exp -2\int_0^\infty ds \int \mathbf{M}(d\omega)(1 - \exp - \langle \omega, g_s \rangle)$$

$$= \exp -2\int \mathbf{N}(d\omega)(1 - \exp - \langle \omega, g \rangle) \qquad \text{from the definition of } \mathbf{N}.$$

Thus, we have completely proved the theorem.

3.3 An extension of the Ray-Knight theorems

(3.3.1) Now that we have obtained the Lévy-Khintchine representation of Q_x^δ, we may use the infinite divisibility property again to obtain some extensions of the basic Ray-Knight theorems.

First of all, it may be of some interest to define squares of Bessel processes with generalized dimensions, that is: some $\mathbb{R}_+$-valued processes which satisfy:

$$(*) \qquad\qquad X_t = x + 2 \int_0^t \sqrt{X_s} d\beta_s + \Delta(t)$$

where $\Delta : \mathbb{R}_+ \to \mathbb{R}_+$ is a strictly increasing, continuous C^1-function, with $\Delta(0) = 0$ and $\Delta(\infty) = \infty$.

Then, it is not difficult to show, with the help of some weak convergence argument, that the law Q_x^Δ of the unique solution of $(*)$ satisfies:

$$Q_x^\Delta(e^{-I_f}) = \exp -\int \mathbf{M}(d\omega) \left\{ x \left(1 - \exp -I_f(\omega)\right) + \int_0^\infty \Delta(ds)(1 - \exp -I_{f_s}(\omega)) \right\}.$$

Now, we have the following

Theorem 3.3 *The family of local times of $\left(|B_u| + \Delta^{-1}(2\ell_u); u \geq 0\right)$ is Q_0^Δ. In particular, the family of local times of $\left(|B_u| + \frac{2}{\delta}\ell_u; u \geq 0\right)$ is Q_0^δ.*

PROOF: We use Proposition 3.1 with

$$M_t = \exp -\int_0^t ds\, f\left(|B_s| + \Delta^{-1}(2\ell_s)\right) \quad,$$

and we obtain, for any $x \geq 0$:

$$E[M_{\tau_x}] = \exp - \int_0^x ds \int \mathbf{n}(d\varepsilon) \left\{ \left(1 - \exp - \int_0^R du\, f\left(|\varepsilon(u)| + \Delta^{-1}(2s)\right)\right) \right\}$$

$$= \exp -2 \int_0^x ds \int \mathbf{n}_+(d\varepsilon) \left\{ \left(1 - \exp - \int_0^R du\, f\left(|\varepsilon(u)| + \Delta^{-1}(2s)\right)\right) \right\}$$

$$= \exp - \int_0^{2x} dt \int \mathbf{n}_+(d\varepsilon) \left\{ \left(1 - \exp - \int_0^R du\, f\left(|\varepsilon(u)| + \Delta^{-1}(t)\right)\right) \right\}$$

$$= \exp - \int_0^{\Delta^{-1}(2x)} d\Delta(h) \int \mathbf{n}_+(d\varepsilon) \left(1 - \exp - \int_0^R du\, f(\varepsilon(u) + h)\right) ,$$

and the result of the theorem now follows by letting $x \to \infty$. $\square$

In fact, in the previous proof, we showed more than the final statement, since we considered the local times of $\left(|B_u| + \Delta^{-1}(2\ell_u) : u \le \tau_x\right)$. In particular, the above proof shows the following

Theorem 3.4 *Let $x > 0$, and consider $\tau_x \equiv \inf\{t \ge 0 : \ell_t > x\}$. Then, the processes $\left(\ell_{\tau_x}^{a-2x/\delta}\left(|B| - \frac{2}{\delta}\ell\right); a \ge 0\right)$ and $\left(\ell_{\tau_x}^a\left(|B| + \frac{2}{\delta}\ell\right); a \ge 0\right)$ have the same law, namely that of an inhomogeneous Markov process $(Y_a; a \ge 0)$, starting at 0, which is the square of a δ-dimensional Bessel process for a $\le \frac{2x}{\delta}$, and a square Bessel process of dimension 0, for $a \ge \frac{2x}{\delta}$.*

(3.3.2) These connections between Brownian occupation times and squares of Bessel processes explain very well why, when computing quantities to do with Brownian occupation times, we find formulae which also appeared in relation with Lévy's formula (see Chapter 2). Here is an important example.

We consider a one-dimensional Brownian motion $(B_t, t \ge 0)$, starting from 0, and we define $\sigma = \inf\{t : B_t = 1\}$, and $S_t = \sup_{s \le t} B_s$ $(t \ge 0)$. Let $a < 1$; we are interested in the joint distribution of the triple:

$$A_\sigma^-(a) \overset{\text{def}}{=} \int_0^\sigma ds\, 1_{(B_s < aS_s)} ; \quad \ell_\sigma^{(a)} \overset{\text{def}}{=} \ell_\sigma^0(B - aS) ; \quad A_\sigma^+(a) \overset{\text{def}}{=} \int_0^\sigma ds\, 1_{(B_s > aS_s)}$$

Using standard stochastic calculus, we obtain: for every $\mu, \lambda, \nu > 0$,

$$E\left[\exp - \left(\frac{\mu^2}{2}A_\sigma^-(a) + \lambda\ell_\sigma^{(a)} + \frac{\nu^2}{2}A_\sigma^+(a)\right)\right] = \left(\mathrm{ch}(\nu\bar{a}) + (\mu + 2\lambda)\frac{\mathrm{sh}(\nu\bar{a})}{\nu}\right)^{-1/\bar{a}}$$

where $\bar{a} = 1 - a > 0$.

On the other hand, we deduce from formula (2.1) and the additivity property, presented in Theorem 2.3, of the family $(Q_x^\delta; \delta \geq 0, x \geq 0)$ the following formula: for every $\delta \geq 0$, and $\nu, \lambda, x \geq 0$,

$$Q_0^\delta \left(\exp - \left(\frac{\nu^2}{2} \int\limits_0^x dy\, X_y + \lambda\, X_x \right) \right) = \left(\mathrm{ch}(\nu x) + \frac{2\lambda}{\nu}\mathrm{sh}(\nu x) \right)^{-\delta/2}$$

Comparing the two previous expectations, we obtain the following identity in law, for $b > 0$:

$$\left(A_\sigma^+(1-b); \ell_\sigma^{(1-b)} \right) \overset{\text{(law)}}{=} \left(\int\limits_0^b dy\, X_y^{(2/b)};\ X_b^{(2/b)} \right) \qquad (*)\,,$$

where, on the right-hand side of $(*)$, we denote by $(X_y^{(\delta)}, y \geq 0)$ a BESQ$^\delta$ process, starting from 0.

Thanks to Lévy's representation of reflecting Brownian motion as $(S_t - B_t; t \geq 0)$, the left-hand side in $(*)$ is identical in law to:

$$\left(\int\limits_0^b dy\, \ell_{T_1}^{y-b} \left(|B| - b\ell^0 \right)\ ;\ \ell_{T_1}^0 \left(|B| - b\ell^0 \right) \right)$$

Until now in this subparagraph (3.3.2), we have not used any Ray-Knight theorem; however, we now do so, as we remark that the identity in law between the last written pair of r.v.'s and the right-hand side of $(*)$ follows directly from Theorem 3.4.

3.4 The law of Brownian local times taken at an independent exponential time

The basic Ray-Knight theorems (RK1) and (RK2) express the laws of Brownian local times in the space variable up to some particular stopping times, namely τ_x and T_1. It is a natural question to look for an identification of the law of Brownian local times up to a fixed time t. One of the inherent difficulties of this question is that now, the variable B_t is not a constant; one way to circumvent this problem would be to condition with respect to the variable B_t; however, even when this is done, the answer to the problem is not particularly simple (see Perkins [68], and Jeulin [52]). In fact, if one considers the same problem at an independent exponentially distributed time, it then turns out that all is needed is to combine the two basic RK theorems. This shows up clearly in the next

Proposition 3.2 *Let S_θ be an independent exponential time, with parameter $\frac{\theta^2}{2}$, that is: $P(S_\theta \in ds) = \frac{\theta^2}{2} \exp\left(-\frac{\theta^2 s}{2}\right) ds$. Then*

1) ℓ_{S_θ} *and* B_{S_θ} *are independent, and have respective distributions:*

$$P(\ell_{S_\theta} \in d\ell) = \theta e^{-\theta \ell} d\ell \quad ; \quad P(B_{S_\theta} \in da) = \frac{\theta}{2} e^{-\theta|a|} da .$$

2) *for any* $\mathbb{R}_+$ *valued, continuous additive functional A, the following formula holds:*

$$E\left[\exp(-A_{S_\theta}) \mid \ell_{S_\theta} = \ell; B_{S_\theta} = a\right]$$
$$= E\left[\exp\left(-A_{T_\ell} - \frac{\theta^2}{2} T_\ell\right)\right] e^{\theta \ell} E_a\left[\exp\left(-A_{T_0} - \frac{\theta^2}{2} T_0\right)\right] e^{\theta|a|}$$

Then, using the same sort of *transfer principle* arguments as we did at the end of paragraph (3.1), one obtains the following

Theorem 3.5 *Conditionally on $\ell^0_{S_\theta} = \ell$, and $B_{S_\theta} = a > 0$, the process $(\ell^x_{S_\theta}; x \in \mathbb{R})$ is an inhomogeneous Markov process which may be described as follows:*

i) $(\ell^{-x}_{S_\theta}; x \geq 0)$ *and* $(\ell^x_{S_\theta}; x \geq a)$ *are diffusions with common infinitesimal generator:*

$$2y \frac{d^2}{dy^2} - 2\theta y \frac{d}{dy}$$

ii) $(\ell^x_{S_\theta}; 0 \leq x \leq a)$ *is a diffusion with infinitesimal generator:*

$$2y \frac{d^2}{dy^2} + (2 - 2\theta y) \frac{d}{dy} .$$

This theorem may be extended to describe the local times of $|B| + \frac{2}{\delta}\ell^0$, considered up to an independent exponential time (see Biane-Yor [19]).

Exercise 3.1 Extend the second statement of Proposition 3.2 by showing that, if A^- and A^+ are two $\mathbb{R}_+$-valued continuous additive functionals, the following formula holds:

$$E\left[\exp - \left(A_{g_{S_\theta}}^- + A_{S_\theta}^+ - A_{g_{S_\theta}}^+\right) \mid \ell_{S_\theta} = \ell; B_{S_\theta} = a\right]$$

$$= E\left[\exp - \left(A_{\tau_\ell}^- + \frac{\theta^2}{2}\tau_\ell\right)\right] e^{\theta\ell} E_a\left[\exp - \left(A_{T_0}^+ + \frac{\theta^2}{2}T_0\right)\right] e^{\theta|a|}$$

3.5 Squares of Bessel processes and squares of Bessel bridges

From the preceding discussion, the reader might draw the conclusion that the extension of Ray-Knight theorems from Brownian (or Bessel) local times to the local times of the processes: $\Sigma_t^\delta \equiv |B_t| + \frac{2}{\delta}\ell_t$ ($t \geq 0$) is plain-sailing.

It will be shown, in this paragraph, that except for the case $\delta = 2$, the non-Markovian character of Σ^δ creates some important, and thought pro-voking, difficulties. On a more positive view point, we present an additive decomposition of the square of a Bessel process of dimension δ as the sum of the square of a δ-dimensional Bessel bridge, and an interesting independent process, which we shall describe. In terms of convolution, we show:

$$(*) \qquad\qquad Q_0^\delta = Q_{0\to 0}^\delta * \underline{R}^\delta \ ,$$

where $\underline{R}^\delta$ is a probability on Ω_+^*, which shall be identified. (We hope that the notation $\underline{R}^\delta$ for this *remainder* or *residual* probability will not create any confusion with the notation for Bessel processes, often written as $(R_\delta(t), t \geq 0)$; the context should help...)

(3.5.1) **The case $\delta = 2$.**

In this case, the decomposition $(*)$ is obtained by writing:

$$\ell_\infty^a(R_3) = \ell_{T_1}^a(R_3) + \left(\ell_\infty^a(R_3) - \ell_{T_1}^a(R_3)\right) \ .$$

The process $\left(\ell_{T_1}^a(R_3); 0 \leq a \leq 1\right)$ has the law $Q_{0 \to 0}^2$, which may be seen by a Markovian argument:

$$\left(\ell_{T_1}^a(R_3); 0 \leq a \leq 1\right) \stackrel{\text{(law)}}{=} \left\{\left(\ell_\infty^a(R_3); 0 \leq a \leq 1\right) \middle| \ell_\infty^1(R_3) = 0\right\}$$

but we shall also present a different argument in the subparagraph (3.5.3).

We now define $\underline{R}^2$ as the law of $\left(\ell_\infty^a(R_3) - \ell_{T_1}^a(R_3); 0 \leq a \leq 1\right)$ which is also, thanks to the strong Markov property of R_3, the law of the local times $\left(\ell_\infty^a(R_3^{(1)}); 0 \leq a \leq 1\right)$ below level 1, of a 3-dimensional Bessel process starting from 1.

In the sequel, we shall use the notation $\hat{P}$ to denote the probability on Ω_+^* obtained by time reversal at time 1 of the probability P, that is:

$$\hat{E}\left[F(X_t; t \leq 1)\right] = E\left[F(X_{1-t}; t \leq 1)\right] \ .$$

We may now state two interesting representations of $\underline{R}^2$. First, $\underline{R}^2$ can be represented as:

$$\underline{R}^2 = \mathcal{L}\left(r_4^2\left((a-U)^+\right); 0 \leq a \leq 1\right) \ , \tag{3.1}$$

where $\mathcal{L}\left(\gamma(a); 0 \leq a \leq 1\right)$ denotes the law of the process γ, and $(r_4(a); 0 \leq a \leq 1)$ denotes a 4-dimensional Bessel process starting from 0, and U is a uniform variable on $[0,1]$, independent of r_4. This representation follows from Williams' path decomposition of Brownian motion $(B_t; t \leq \sigma)$, where $\sigma = \inf\{t : B_t = 1\}$, and (RK2).

The following representation of $\underline{R}^2$ is also interesting:

$$\underline{R}^2 = \int_0^\infty \frac{dx}{2} e^{-x/2} \hat{Q}_{x \to 0}^0 \tag{3.2}$$

This formula may be interpreted as:

the law of $\left(\ell_\infty^a(R_3^{(1)}); 0 \leq a \leq 1\right)$ given $\ell_\infty^1(R_3^{(1)}) = x$ is $\hat{Q}_{x \to 0}^0$

or, using Williams' time reversal result:

$$\hat{\underline{R}}^2 = \int_0^\infty \frac{dx}{2} e^{-x/2} Q_{x \to 0}^0 \ \text{is the law of} \ \left(\ell_{g_\theta}^a(B); 0 \leq a \leq 1\right) \ , \tag{3.3}$$

where $g_\sigma = \sup\{t \leq \sigma : B_t = 0\}$.

To prove (3.2), we condition Q_0^2 with respect to X_1, and we use the additivity and time reversal properties of the squared Bessel bridges. More precisely, we have:

$$Q_0^2 = \int_0^\infty \frac{dx}{2} e^{-x/2} Q_{0 \to x}^2 = \int_0^\infty \frac{dx}{2} e^{-x/2} \hat{Q}_{x \to 0}^2 .$$

However, we have: $Q_{x \to 0}^2 = Q_{0 \to 0}^2 * Q_{x \to 0}^0$, hence:

$$\hat{Q}_{x \to 0}^2 = Q_{0 \to 0}^2 * \hat{Q}_{x \to 0}^0 ,$$

so that we now obtain:

$$Q_0^2 = Q_{0 \to 0}^2 * \int_0^\infty \frac{dx}{2} e^{-x/2} \hat{Q}_{x \to 0}^0$$

Comparing this formula with the definition of $\underline{R}^2$ given in $(*)$, we obtain (3.2).

(3.5.2) The general case $\delta > 0$.

Again, we decompose Q_0^δ by conditioning with respect to X_1, and using the additivity and time reversal properties of the squared Bessel bridges. Thus, we have:

$$Q_0^\delta = \int_0^\infty \gamma_\delta(dx) Q_{0 \to x}^\delta , \quad \text{where} \quad \gamma_\delta(dx) = Q_0^\delta(X_1 \in dx) = \frac{dx}{2} \left(\frac{x}{2}\right)^{\frac{\delta}{2}-1} \frac{e^{-x/2}}{\Gamma\left(\frac{\delta}{2}\right)}.$$

From the additivity property:

$$Q_{x \to 0}^\delta = Q_{0 \to 0}^\delta * Q_{x \to 0}^0 ,$$

we deduce:

$$Q_{0 \to x}^\delta = Q_{0 \to 0}^\delta * \hat{Q}_{x \to 0}^0$$

and it follows that:

$$Q_0^\delta = Q_{0 \to 0}^\delta * \int_0^\infty \gamma_\delta(dx) \hat{Q}_{x \to 0}^0 ,$$

so that:

$$\underline{R}^\delta = \int_0^\infty \gamma_\delta(dx) \hat{Q}_{x \to 0}^0 \equiv \int_0^\infty \gamma_2(dx) g_\delta(x) \hat{Q}_{x \to 0}^0 ,$$

with:

$$g_\delta(x) = c_\delta x^{\frac{\delta}{2}-1} \quad , \quad \text{where } c_\delta = \frac{1}{\Gamma\left(\frac{\delta}{2}\right)} \frac{1}{2^{\frac{\delta}{2}-1}} \; .$$

Hence, we have obtained the following relation:

$$\underline{R}^\delta = c_\delta(X_1)^{\frac{\delta}{2}-1}\underline{R}^2 \; ,$$

and we may state the following

Theorem 3.6 *For any $\delta > 0$, the additive decomposition:*

$$Q_0^\delta = Q_{0\to 0}^\delta * \underline{R}^\delta$$

holds, where $\underline{R}^\delta$ may be described as follows:
$\underline{R}^\delta$ is the law of the local times, for levels $a \leq 1$, of the 3-dimensional Bessel process, starting from 1, with weight: $c_\delta(\ell_\infty^1(R_3))^{\frac{\delta}{2}-1}$, or, equivalently:
$\hat{\underline{R}}^\delta$ is the law of the local times process: $(\ell_{g_\sigma}^a(B^{(\delta)}); 0 \leq a \leq 1)$
where $B^{(\delta)}$ has the law W^δ defined by:

$$W^\delta_{|\mathcal{F}_\sigma} = c_\delta(\ell_\sigma^0)^{\frac{\delta}{2}-1} \cdot W_{|\mathcal{F}_\sigma}$$

Before going any further, we remark that the family $(\underline{R}^\delta, \delta > 0)$ also possesses the additivity property:

$$\underline{R}^{\delta+\delta'} = \underline{R}^\delta * \underline{R}^{\delta'}$$

and, with the help of the last written interpretation of $\underline{R}^\delta$, we can now present the following interesting formula:

Theorem 3.7 *Let $f : \mathbb{R} \to \mathbb{R}_+$ be any Borel function. Then we have:*

$$W^\delta\left(\exp - \int_0^{g_\sigma} ds \, f(B_s)\right) = \left(W\left(\exp - \int_0^{g_\sigma} ds \, f(B_s)\right)\right)^{\delta/2}$$

(3.5.3)   **An interpretation of $Q_{0\to 0}^\delta$**

The development presented in this subparagraph follows from the well-known fact that, if $(b(t); 0 \leq t \leq 1)$ is a standard Brownian bridge, then:

(*) $B_t = (t+1)b\left(\dfrac{t}{t+1}\right) \quad , \quad t \geq 0 \; ,$

is a Brownian motion starting from 0, and, conversely, the formula $(*)$ allows to define a Brownian bridge b from a Brownian motion B.

Consequently, to any Borel function $\tilde{f} : [0,1] \to \mathbb{R}_+$, there corresponds a Borel function $f : \mathbb{R}_+ \to \mathbb{R}_+$, and conversely, such that:

$$\int_0^\infty dt \, f(t) B_t^2 = \int_0^1 du \, \tilde{f}(u) b^2(u) \ .$$

This correspondance is expressed explicitely by the two formulae:

$$f(t) = \frac{1}{(1+t)^4} \tilde{f}\left(\frac{t}{t+1}\right) \quad \text{and} \quad \tilde{f}(u) = \frac{1}{(1+u)^4} f\left(\frac{u}{1-u}\right)$$

These formulae, together with the additivity properties of Q_0^δ and $Q_{0\to0}^\delta$ allow us to obtain the following

Theorem 3.8 *Define* $\left(D_t^\delta, t < \tilde{T}_1^\delta \equiv \displaystyle\int_0^\infty \frac{ds}{(1+\Sigma_s^\delta)^4}\right)$ *via the following space and time change formula:*

$$\frac{\Sigma_t^\delta}{1+\Sigma_t^\delta} = D^\delta\left(\int_0^t \frac{ds}{(1+\Sigma_s^\delta)^4}\right)$$

Then, $Q_{0\to0}^\delta$ *is the law of the local times of* $(D_t^\delta, t < \tilde{T}_1^\delta)$.

(Remark that $(D_t^\delta, t < \tilde{T}_1^\delta)$ may be extended by continuity to $t = \tilde{T}_1^\delta$, and then we have: $\tilde{T}_1^\delta = \inf\{t : D_t^\delta = 1\}$.).

PROOF: For any Borel function $\tilde{f} : [0,1] \to \mathbb{R}_+$, we have, thanks to the remarks made previously:

$$Q_{0\to 0}^{\delta}\left(\exp-\langle\omega,\tilde{f}\rangle\right) = Q_0^{\delta}\left(\exp-\langle\omega,f\rangle\right)$$

$$= E\left[\exp\left(-\int_0^{\infty} du\, f\left(\Sigma_u^{\delta}\right)\right)\right] \qquad \text{(from Theorem (3.3))}$$

$$= E\left[\exp-\int_0^{\infty}\frac{du}{(1+\Sigma_u^{\delta})^4}\tilde{f}\left(\frac{\Sigma_u^{\delta}}{1+\Sigma_u^{\delta}}\right)\right] \qquad \text{(from the relation } f\leftrightarrow\tilde{f})$$

$$= E\left[\exp-\int_0^{\tilde{T}_1^{\delta}} dv\,\tilde{f}(D_v^{\delta})\right] \qquad \text{(from the definition of } D^{\delta})$$

The theorem is proven. □

It is interesting to consider again the case $\delta = 2$ since, as argued in (3.5.1), it is then known that $Q_{0\to 0}^2$ is the law of the local times of $(R_3(t), t \leq T_1(R_3))$. This is perfectly coherent with the above theorem, since we then have:

$$(D_t^2; t \leq T_1^2) \overset{\text{(law)}}{=} (R_3(t), t \leq T_1(R_3))$$

PROOF: If we define: $X_t = \dfrac{R_3(t)}{1+R_3(t)}$ $(t \geq 0)$, we then have:

$$\frac{1}{X_t} = 1 + \frac{1}{R_3(t)} \quad ;$$

therefore, $(X_t, t \geq 0)$, which is a diffusion (from its definition in terms of R_3) is also such that $\left(\frac{1}{X_t}, t \geq 0\right)$ is a local martingale. Then, it follows easily that $X_t = \tilde{R}_3(\langle X\rangle_t), t \geq 0$, where $(\tilde{R}_3(u), u \geq 0)$ is a 3 dimensional Bessel process, and, finally, since:

$$\langle X\rangle_t = \int_0^t \frac{ds}{(1+R_3(s))^4}, \text{ we get the desired result.}$$

Remark: We could have obtained this result more directly by applying Itô's formula to $g(r) = \frac{r}{1+r}$, and then time-changing. But, in our opinion, the above proof gives a better explanation of the ubiquity of R_3 in this question.

Exercise 3.2:   Let $a, b > 0$, and $\delta > 2$. Prove that, if $(R_\delta(t), t \geq 0)$ is a δ-dimensional Bessel process, then: $R_\delta(t)/(a + b(R_\delta(t))^{\delta-2})^{1/\delta-2}$ may be obtained by time changing a δ-dimensional Bessel process, up to its first hitting time of $c = b^{-(1/\delta-2)}$.

This exercise may be generalised as follows:

Exercise 3.3: Let $(X_t, t \geq 0)$ be a real-valued diffusion, whose infinitesimal generator L satisfies:

$$L\varphi(x) = \frac{1}{2}\varphi''(x) + b(x)\varphi'(x) , \quad \text{for } \varphi \in C^2(\mathbb{R}) .$$

Let $f : \mathbb{R} \to \mathbb{R}_+$ be a C^2 function, and, finally, let $\delta > 1$.

1. Prove that, if b and f are related by:

$$b(x) = \frac{\delta - 1}{2}\frac{f'(x)}{f(x)} - \frac{1}{2}\frac{f''(x)}{f'(x)}$$

then, there exists $(R_\delta(u), u \geq 0)$ a δ-dimensional Bessel process, possibly defined on an enlarged probability space, such that:

$$f(X_t) = R_\delta\left(\int_0^t ds(f')^2(X_s)\right) , \quad t \geq 0 .$$

2. Compute $b(x)$ in the following cases:

$$\text{(i)} \ \ f(x) = x^\alpha ; \quad \text{(ii)} \ \ f(x) = \exp(ax) .$$

3.6 Generalized meanders and squares of Bessel processes

(3.6.1)   The Brownian meander, which plays an important role in a number of studies of Brownian motion, may be defined as follows:

$$m(u) = \frac{1}{\sqrt{1-g}}|B_{g+u(1-g)}| , \quad u \leq 1 ,$$

where $g = \sup\{u \leq 1 : B_u = 0\}$, and $(B_t, t \geq 0)$ denotes a real-valued Brownian motion starting from 0.

Imhof [49] proved the following absolute continuity relation:

$$M = \frac{c}{X_1} \cdot S \qquad \left(c = \sqrt{\frac{\pi}{2}}\right) \qquad\qquad (3.4)$$

where M, resp.: S, denotes the law of $(m(u), u \leq 1)$, resp.: the law of $(R(u), u \leq 1)$ a BES(3) process, starting from 0.

Other proofs of (3.4) have been given by Biane-Yor [18], using excursion theory, and Azéma-Yor ([1], paragraph 4) using an extension of Girsanov theorem.

It is not difficult, using the same kind of arguments, to prove the more general absolute continuity relationship:

$$M_\nu = \frac{c_\nu}{X_1^{2\nu}} \cdot S_\nu \qquad\qquad (3.5)$$

where $\nu \in (0,1)$ and M_ν, resp.: S_ν, denotes the law of

$$m_\nu(u) \equiv \frac{1}{\sqrt{1 - g_\nu}} R_{-\nu}(g_\nu + u(1 - g_\nu)) \qquad (u \leq 1)$$

the Bessel meander associated to the Bessel process $R_{-\nu}$ of dimension $2(1-\nu)$, starting from 0, resp.: the law of the Bessel process of dimension $2(1 + \nu)$, starting from 0.

Exercise 3.4: Deduce from formula (3.5) that:

$$M_\nu [F(X_u; u \leq 1) \mid X_1 = x] = S_\nu [F(X_u; u \leq 1) \mid X_1 = x]$$

and that:

$$M_\nu(X_1 \in dx) = x \exp\left(-\frac{x^2}{2}\right) dx .$$

In particular, the law of $m_\nu(1)$, the value at time 1 of the Bessel meander does not depend on ν, and is distributed as the 2-dimensional Bessel process at time 1 (See Corollary 3.6.1 for an explanation).

(3.6.2) Biane-Le Gall-Yor [16] proved the following absolute continuity relation, which looks similar to (3.5): for every $\nu > 0$,

$$N_\nu = \frac{2\nu}{X_1^2} \cdot S_\nu \qquad\qquad (3.6)$$

where N_ν denotes the law on $C([0,1], \mathbb{R}_+)$ of the process:

$$n_\nu(u) = \frac{1}{\sqrt{\mathcal{L}^\nu}} R_\nu(\mathcal{L}^\nu u) \qquad (u \leq 1) ,$$

with $\mathcal{L}^\nu \equiv \sup\{t > 0 : R_\nu(t) = 1\}$.

Exercise 3.5: Deduce from formula (3.6) that:

$$N_\nu \left[F(X_u; u \le 1) \mid X_1 = x \right] = S_\nu \left[F(X_u; u \le 1) \mid X_1 = x \right]$$

and that:

$$N_\nu(X_1 \in dx) = S_{\nu-1}(X_1 \in dx) \ .$$

(Corollary 3.9.2 gives an explanation of this fact)

(3.6.3) In this sub-paragraph, we shall consider, more generally than the right-hand sides of (3.5) and (3.6), the law S_ν modified via a Radon-Nikodym density of the form:

$$\frac{c_{\mu,\nu}}{X_1^\mu} \ ,$$

and we shall represent the new probability in terms of the laws of Bessel processes and Bessel bridges.

Precisely, consider $(R_t, t \le 1)$ and $(R'_t, t \le 1)$ two independent Bessel processes, starting from 0, with respective dimensions d and d'; condition R by $R_1 = 0$, and define $M^{d,d'}$ to be the law of the process $\left((R_t^2 + (R'_t)^2)^{1/2}, t \le 1 \right)$ obtained in this way; in other terms, the law of the square of this process, that is: $\left(R_t^2 + (R'_t)^2, t \le 1 \right)$ is $Q^d_{0 \to 0} * Q^{d'}_0$.

We may now state and prove the following

Theorem 3.9 *Let P_0^δ be the law on $C([0,1]; \mathbb{R}_+)$ of the Bessel process with dimension δ, starting from 0. Then, we have:*

$$M^{d,d'} = \frac{c_{d,d'}}{X_1^d} \cdot P_0^{d+d'} \tag{3.7}$$

where $c_{d,d'} = M^{d,d'}(X_1^d) = (2^{d'/2}) \dfrac{\Gamma\left(\frac{d+d'}{2}\right)}{\Gamma\left(\frac{d}{2}\right)}.$

PROOF: From the additivity property of squares of Bessel processes, which in terms of the probabilities $(Q_x^\delta; \delta \ge 0, x \ge 0)$ is expressed by:

$$Q_x^d * Q_{x'}^{d'} = Q_{x+x'}^{d+d'}$$

(see Theorem 2.3 above), it is easily deduced that:

$$Q_{x+x' \to 0}^{d+d'} = Q_{x \to 0}^d * Q_{x' \to 0}^{d'} \ .$$

Hence, by reverting time from $t = 1$, we obtain:

$$Q_{0 \to x+x'}^{d+d'} = Q_{0 \to x}^{d} * Q_{0 \to x'}^{d'} \; ,$$

and, in particular:

$$Q_{0 \to x}^{d+d'} = Q_{0 \to 0}^{d} * Q_{0 \to x}^{d'} \; .$$

From this last formula, we deduce that, conditionally on $X_1 = x$, both sides of (3.7) are equal, so that, to prove the identity completely, it remains to verify that the laws of X_1 relatively to each side of (3.7) are the same, which is immediate. □

As a consequence of Theorem 3.9, and of the absolute continuity relations (3.5) and (3.6), we are now able to identify the laws $M_\nu (0 < \nu < 1)$, and N_ν ($\nu > 0$), as particular cases of $M^{d,d'}$.

Corollary 3.9.1   *Let $0 < \nu < 1$. Then, we have:*

$$M_\nu = M^{2\nu,2}$$

In other words, the square of the Bessel meander of dimension $2(1 - \nu)$ may be represented as the sum of the squares of a Bessel bridge of dimension 2ν and of an independent two-dimensional Bessel process.

In the particular case $\nu = 1/2$, the square of the Brownian meander is distributed as the sum of the squares of a Brownian bridge and of an independent two-dimensional Bessel process.

Corollary 3.9.2   *Let $\nu > 0$. Then, we have:*

$$N_\nu = M^{2,2\nu}$$

In other words, the square of the normalized Bessel process $\left(\frac{1}{\sqrt{\mathcal{L}^\nu}} R(\mathcal{L}^\nu u); \leq 1 \right)$ with dimension $d = 2(1 + \nu)$ is distributed as the sum of the squares of a two dimensional Bessel bridge and of an independent Bessel process of dimension 2ν.

3.7 Generalized meanders and Bessel bridges

(3.7.1)   As a complement to the previous paragraph 3.6, we now give a representation of the Bessel meander m_ν (defined just below formula (3.5)) in terms of the Bessel bridge of dimension $\delta_\nu^+ \equiv 2(1+\nu)$.

We recall that this Bessel bridge may be realized as:

$$\rho_\nu(u) = \frac{1}{\sqrt{d_\nu - g_\nu}} R_{-\nu}\left(g_\nu + u(d_\nu - g_\nu)\right) \quad , \qquad u \le 1,$$

where $d_\nu = \inf\{u \ge 1 : R_{-\nu}(u) = 0\}$.

Comparing the formulae which define m_ν and ρ_ν, we obtain

Theorem 3.10 *The following equality holds:*

$$m_\nu(u) = \frac{1}{\sqrt{V_\nu}} \rho_\nu(uV_\nu) \qquad (u \le 1) \tag{3.8}$$

where $V_\nu = \dfrac{1 - g_\nu}{d_\nu - g_\nu}$.

Furthermore, V_ν and the Bessel bridge $(\rho_\nu(u), u \le 1)$ are independent, and the law of V_ν is given by:

$$P(V_\nu \in dt) = \nu t^{\nu - 1} dt \qquad (0 < t < 1) \ .$$

Similarly, it is possible to present a realization of the process n_ν in terms of ρ_ν.

Theorem 3.11 *1) Define the process:*

$$\tilde{n}_\nu(u) = \frac{1}{\sqrt{d_\nu - 1}} \rho_\nu(d_\nu - u(d_\nu - 1)) \equiv \frac{1}{\sqrt{\hat{V}_\nu}} \hat{\rho}_\nu(u\hat{V}_\nu) \qquad (u \le 1)$$

where

$$\hat{\rho}_\nu(u) = \rho_\nu(1 - u) \quad (u \le 1) \ and \ \hat{V}_\nu = 1 - V_\nu = \frac{d_\nu - 1}{d_\nu - g_\nu} \ .$$

Then, the processes n_ν and $\tilde{n}_\nu$ have the same distribution.

2) Consequently, the identity in law

$$(n_\nu(u), u \leq 1) \overset{\text{(law)}}{=} \left(\frac{1}{\sqrt{\hat{V}_\nu}} \hat{\rho}_\nu(u\hat{V}_\nu), u \leq 1 \right) \tag{3.9}$$

holds, where, on the right-hand side, $\hat{\rho}_\nu$ is a Bessel bridge of dimension $\delta_\nu^+ \equiv 2(1 + \nu)$, and $\hat{V}_\nu$ is independent of $\hat{\rho}_\nu$, with:

$$P(\hat{V}_\nu \in dt) = \nu(1 - t)^{\nu - 1} dt \qquad (0 < t < 1) \ .$$

(3.7.2)   Now, the representations of the processes m_ν and n_ν given in Theorems 3.10 and 3.11 may be generalized as follows to obtain a representation of a process whose distribution is $M^{d,d'}$ (see Theorem 3.9 above).

Theorem 3.12 *Let $d, d' > 0$, and define $(\rho^{d+d'}(u), u \leq 1)$ to be the Bessel bridge with dimension $d + d'$.*

Consider, moreover, a beta variable $V_{d,d'}$, with parameters $\left(\frac{d}{2}, \frac{d'}{2} \right)$, i.e.:

$$P(V_{d,d'} \in dt) = \frac{t^{\frac{d}{2}-1}(1 - t)^{\frac{d'}{2}-1} dt}{B\left(\frac{d}{2}, \frac{d'}{2} \right)} \qquad (0 < t < 1)$$

such that $V_{d,d'}$ is independent of $\rho^{d+d'}$.

Then, the distribution of the process:

$$\left(m^{d,d'}(u) \overset{\text{def}}{=} \frac{1}{\sqrt{V_{d,d'}}} \rho^{d+d'}(uV_{d,d'}), u \leq 1 \right)$$

is $M^{d,d'}$.

In order to prove Theorem 3.12, we shall use the following Proposition, which relates the laws of the Bessel bridge and Bessel process, for any dimension.

Proposition 3.3 *Let Π_μ, resp.: S_μ, be the law of the standard Bessel bridge, resp.: Bessel process, starting from 0, with dimension $\delta = 2(\mu + 1)$.*

Then, for any $t < 1$ and every Borel functional $F : C([0, t], \mathbb{R}_+) \to \mathbb{R}_+$, we have:

$$\Pi_\mu[F(X_u, u \le t)] = S_\mu\left[F(X_u, u \le t)h_\mu(t, X_t)\right]$$

where:

$$h_\mu(t, x) = \frac{1}{(1-t)^{\mu+1}} \exp{-\frac{x^2}{2(1-t)}}$$

PROOF: This is a special case of the partial absolute continuity relationship between the laws of a nice Markov process and its bridges (see, e.g. [41]). □

We now prove Theorem 3.12.

In order to present the proof in a natural way, we look for V, a random variable taking its values in $(0, 1)$, and such that:

i) $P(V \in dt) = \theta(t)dt$;   ii) V is independent of $\rho^{d+d'}$;

iii) the law of the process $\left(\frac{1}{\sqrt{V}}\rho^{d+d'}(uV); u \le 1\right)$ is $M^{d,d'}$.

We define the index μ by the formula: $d + d' = 2(\mu + 1)$. Then, we have, for every Borel function $F : C([0, 1], \mathbb{R}_+) \to \mathbb{R}_+$:

$$E\left[F\left(\frac{1}{\sqrt{V}}\rho^{d+d'}(uV); u \le 1\right)\right] = \int_0^1 dt\theta(t)\Pi_\mu\left[F\left(\frac{1}{\sqrt{t}}X_{ut}; u \le 1\right)\right]$$

(by Proposition 3.3)
$$= \int_0^1 dt\theta(t)S_\mu\left[F\left(\frac{1}{\sqrt{t}}X_{ut}; u \le 1\right)h_\mu(t, X_t)\right]$$

(by scaling)
$$= \int_0^1 dt\theta(t)S_\mu\left[F(X_u; u \le 1)h_\mu(t, \sqrt{t}X_1)\right]$$

$$= S_\mu\left[F(X_u, u \le 1)\int_0^1 dt\theta(t)h_\mu(t, \sqrt{t}X_1)\right]$$

Hence, by Theorem 3.9, the problem is now reduced to finding a function θ such that:

$$\int_0^1 dt\theta(t)h_\mu(t, \sqrt{t}x) = \frac{c_{d,d'}}{x^d}$$

Using the explicit formula for h_μ given in Proposition 3.3, and making some elementary changes of variables, it is easily found, by injectivity of the Laplace transform, that:

$$\theta(t) = \frac{t^{\frac{d}{2}-1}(1-t)^{\frac{d'}{2}-1}}{B\left(\frac{d}{2}, \frac{d'}{2}\right)} \qquad (0 < t < 1)$$

which ends the proof of Theorem 3.12.

(3.7.3) We now end up this Chapter by giving the explicit semimartingale decomposition of the process $m^{d,d'}$, for $d + d' \geq 2$, which may be helpful, at least in particular cases, e.g.: for the processes m_ν and n_ν).

Exercise 3.6: (We retain the notation of Theorem 3.9).

1) Define the process

$$D_u = E_0^{d+d'}\left[\frac{c_{d,d'}}{X_1^d}\bigg|\mathcal{F}_u\right] \qquad (u < 1) \ .$$

Then, prove that:

$$D_u = \frac{1}{(1-u)^{d/2}}\Phi\left(\frac{d}{2}, \frac{d+d'}{2}; -\frac{X_u^2}{2(1-u)}\right) \ ,$$

where $\Phi(a, b; z)$ denotes the confluent hypergeometric function with parameters (a, b) (see Lebedev [63], p.260–268).

Hint: Use the integral representation formula (with $d + d' = 2(1 + \mu)$)

$$\Phi\left(\frac{d}{2}, \frac{d+d'}{2}; -b\right) = c_{d,d'}\int\limits_0^\infty \frac{dt}{(\sqrt{2t})^{d/2}}\left(\frac{t}{b}\right)^{\frac{\mu}{2}}\exp(-(b+t))I_\mu(2\sqrt{bt})$$

$$(3.10)$$

(see Lebedev [63], p. 278) and prove that the right-hand side of formula (3.10) is equal to:

$$E_a^{d,d'}\left(\frac{c_{d,d'}}{X_1^d}\right) \ , \qquad \text{where } a = \sqrt{2b} \ .$$

2) Prove that, under $M^{d,d'}$, the canonical process $(X_u, u \leq 1)$ admits the semimartingale decomposition:

$$X_u = \beta_u + \frac{(d+d')-1}{2}\int_0^u \frac{ds}{X_s} - \int_0^u \frac{dsX_s}{1-s}\left(\frac{\Phi'}{\Phi}\right)\left(\frac{d}{2},\frac{d+d'}{2};-\frac{X_s^2}{2(1-s)}\right)$$

where $(\beta_u, u \leq 1)$ denotes a Brownian motion, and, to simplify the formula, we have written $\frac{\Phi'}{\Phi}(a,b;z)$ for $\frac{d}{dz}(\log \Phi(a,b;z))$.

Comments on Chapter 3

The basic Ray-Knight theorems are recalled in paragraph 3.1, and an easy example of the *transfer principle* is given there.

Paragraph 3.2 is taken from Pitman-Yor [73], and, in turn, following Le Gall-Yor [60], some important extensions (Theorem 3.4) of the RK theorems are given. For more extensions of the RK theorems, see Eisenbaum [39] and Vallois [88].

An illustration of Theorem 3.4, which occurred naturally in the asymptotic study of the windings of the 3-dimensional Brownian motion around certain curves (Le Gall-Yor [62]) is developed in the subparagraph (3.3.2).

There is no easy formulation of a Ray-Knight theorem for Brownian local times taken at a fixed time t (see Perkins [68] and Jeulin [52], who have independently obtained a semimartingale decomposition of the local times in the space variable); the situation is much easier when the fixed time is replaced by an independent exponential time, as is explained briefly in paragraph 3.4, following Biane-Yor [19]; the original result is due to Ray [80], but it is presented in a very different form than Theorem 3.5 in the present chapter.

In paragraphs 3.5, 3.6 and 3.7, some relations between Bessel processes, Bessel bridges and Bessel meanders are presented. In the literature, one will find this kind of study made essentially in relation with Brownian motion and the 3-dimensional Bessel process (see Biane-Yor [18] and Bertoin-Pitman [11] for an exposition of known and new results up to 1994). The extensions which are presented here seem very natural and in the spirit of the first half of Chapter 3, in which the laws of squares of Bessel processes of any dimension are obtained as the laws of certain local times processes.

The discussion in subparagraph (3.5.5), leading to Theorem 3.8, was inspired by Knight [58].

Chapter 4

An explanation and some extensions of the Ciesielski-Taylor identities

The Ciesielski-Taylor identities in law, which we shall study in this Chapter, were published in 1962, that is one year before the publication of the papers of Ray and Knight (1963; [80] and [57]) on Brownian local times; as we shall see below, this is more than a mere coincidence!

Here are these identities: if $(R_\delta(t), t \geq 0)$ denotes the Bessel process of dimension $\delta > 0$, starting at 0, then:

$$\int_0^\infty ds \, 1_{(R_{\delta+2}(s) \leq 1)} \overset{\text{(law)}}{=} T_1(R_\delta) \ , \qquad (4.1)$$

where $T_1(R_\delta) = \inf \{t : R_\delta(t) = 1\}$.

(More generally, throughout this chapter, the notation $H(R_\delta)$ shall indicate the quantity H taken with respect to R_δ).

Except in the case $\delta = 1$, there exists no path decomposition explanation of (4.1); in this Chapter, a spectral type explanation shall be provided, which relies essentially upon the two following ingredients:

a) both sides of (4.1) may be written as integrals with respect to the Lebesgue measure on $[0, 1]$ of the total local times of $R_{\delta+2}$ for the left-hand side, and of R_δ, up to time $T_1(R_\delta)$, for the right-hand side; moreover, the laws of the two local times processes can be deduced from (RK2)(a) (see Chapter 3, paragraph 1);

b) the use of the integration by parts formula obtained in Chapter 2, Theorem 2.1.

This method (the use of ingredient b) in particular) allows to extend the identities (4.1) by considering the time spent in an annulus by a $(\delta + 2)$ dimensional Brownian motion; they may also be extended to pairs of diffusions $(X, \hat{X})$ which are much more general than the pairs $(R_{\delta+2}, R_{\delta})$; this type of generalization was first obtained by Ph. Biane [12], who used the expression of the Laplace transforms of the occupation times in terms of differential equations, involving the speed measures and scale functions of the diffusions.

4.1 A pathwise explanation of (4.1) for $\delta = 1$

Thanks to the time-reversal result of D. Williams:

$$(R_3(t), t \leq \mathcal{L}_1(R_3)) \overset{(\text{law})}{=} (1 - B_{\sigma-t}, t \leq \sigma) \ ,$$

where

$$\mathcal{L}_1(R_3) = \sup\{t : R_3(t) = 1\} \ ,$$

and

$$\sigma = \inf\{t : B_t = 1\} \ ,$$

the left-hand side of (4.1) may be written as: $\displaystyle\int_0^{\sigma} ds \, 1_{(B_s > 0)}$, so that, to explain (4.1) in this case, it now remains to show:

$$\int_0^{\sigma} ds \, 1_{(B_s > 0)} \overset{(\text{law})}{=} T_1(|B|) \ . \tag{4.2}$$

To do this, we use the fact that $(B_t^+, t \geq 0)$ may be written as:

$$B_t^+ = |\beta| \left(\int_0^t ds \, 1_{(B_s > 0)} \right) \ , \qquad t \geq 0 \ ,$$

where $(\beta_u, u \geq 0)$ is another one-dimensional Brownian motion starting from 0, so that, from this representation of B^+, we deduce:

$$\int_0^{\sigma} ds \, 1_{(B_s > 0)} = T_1(|\beta|) \ ;$$

this implies (4.2).

4.2 A reduction of (4.1) to an identity in law between two Brownian quadratic functionals

To explain the result for every $\delta > 0$, we write the two members of the C-T identity (4.1) as local times integrals, i.e.

$$\int_0^1 da\, \ell_\infty^a(R_{\delta+2}) \overset{(\text{law})}{=} \int_0^1 da\, \ell_{T_1}^a(R_\delta) \ ,$$

with the understanding that the local times $(\ell_t^a(R_\gamma); a > 0, t \geq 0)$ satisfy the occupation density formula: for every positive measurable f,

$$\int_0^t ds\, f(R_\gamma(s)) = \int_0^\infty da\, f(a)\ell_t^a(R_\gamma) \ .$$

It is not difficult (e.g. see Yor [101]) to obtain the following representations of the local times processes of R_γ taken at $t = \infty$, or $t = T_1(R_\delta)$, with the help of the basic Ray-Knight theorems (see Chapter 3).

Theorem 4.1 Let $(B_t, t \geq 0)$, resp.: $(\tilde{B}_t, t \geq 0)$ denote a planar BM starting at 0, resp. a standard complex Brownian bridge. Then, we have:

1) for $\gamma > 0$, $\quad (\ell_\infty^a(R_{2+\gamma}); a > 0) \overset{(\text{law})}{=} \left(\frac{1}{\gamma a^{\gamma-1}} |B_{a^\gamma}|^2; a > 0 \right)$

$\qquad\qquad\qquad (\ell_{T_1}^a(R_{2+\gamma}); 0 < a \leq 1) \overset{(\text{law})}{=} \left(\frac{1}{\gamma a^{\gamma-1}} |\tilde{B}_{a^\gamma}|^2; 0 < a \leq 1 \right)$

2) for $\gamma = 0$, $\quad (\ell_{T_1}^a(R_2); 0 < a \leq 1) \overset{(\text{law})}{=} \left(a \left| B_{\log\left(\frac{1}{a}\right)} \right|^2; 0 < a \leq 1 \right)$

3) for $0 < \gamma \leq 2$, $\quad (\ell_{T_1}^a(R_{2-\gamma}); 0 < a \leq 1) \overset{(\text{law})}{=} \left(\frac{1}{\gamma a^{\gamma-1}} |B_{1-a^\gamma}|^2; 0 < a \leq 1 \right)$

With the help of this theorem, we remark that the C-T identities (4.1) are equivalent to

$$\frac{1}{\delta} \int_0^1 \frac{da}{a^{\delta-1}} |B_{a^\delta}|^2 \overset{(\text{law})}{=} \frac{1}{\delta-2} \int_0^1 \frac{da}{a^{\delta-3}} |\tilde{B}_{a^{\delta-2}}|^2 \qquad (\delta > 2) \qquad (4.3)$$

$$\frac{1}{2}\int_0^1 \frac{da}{a}|B_{a^2}|^2 \overset{\text{(law)}}{=} \int_0^1 da\, a\left|B_{\left(\log\frac{1}{a}\right)}\right|^2 \qquad (\delta = 2) \qquad (4.4)$$

$$\frac{1}{\delta}\int_0^1 \frac{da}{a^{\delta-1}}|B_{a^\delta}|^2 \overset{\text{(law)}}{=} \int_0^1 \frac{da}{a^{1-\delta}}|B_{1-a^{2-\delta}}|^2 \qquad (\delta < 2) \qquad (4.5)$$

where, on both sides, B, resp. $\tilde{B}$, denotes a complex valued Brownian motion, resp.: Brownian bridge.

In order to prove these identities, it obviously suffices to take real-valued processes for B and $\tilde{B}$, which is what we now assume.

It then suffices to remark that the identity in law (4.4) is a particular case of the integration by parts formula (2.14), considered with $f(a) = \log\frac{1}{a}$, and $g(a) = a^2$; the same argument applies to the identity in law (4.5), with $f(a) = 1 - a^{2-\delta}$, and $g(a) = a^\delta$; with a little more work, one also obtains the identity in law (4.3).

4.3 Some extensions of the Ciesielski-Taylor identities

(4.3.1) The proof of the C-T identities which was just given in paragraph 4.2 uses, apart from the Ray-Knight theorems for (Bessel) local times, the integration by parts formula (obtained in Chapter 2) applied to some functions f and g which satisfy the boundary conditions: $f(1) = 0$ and $g(0) = 0$.

In fact, it is possible to take some more advantage of the integration by parts formula, in which we shall now assume no boundary condition, in order to obtain the following extensions of the C-T identities.

Theorem 4.2 Let $\delta > 0$, and $a \leq b \leq c$. Then, if R_δ and $R_{\delta+2}$ denote two Bessel processes starting from 0, with respective dimensions δ and $\delta + 2$, we have:

$$I_{a,b,c}^{(\delta)} : \quad \begin{aligned} & \int\limits_0^\infty ds\, 1_{(a \le R_{\delta+2}(s) \le b)} + \left(b^{\delta-1} \int\limits_b^c \frac{dx}{x^{\delta-1}} \right) \ell_\infty^b(R_{\delta+2}) \\ & \overset{\text{(law)}}{=} \frac{a}{\delta} \ell_{T_c}^a(R_\delta) + \int\limits_0^{T_c} ds\, 1_{(a \le R_\delta(s) \le b)} \ . \end{aligned}$$

(4.3.2)   We shall now look at some particular cases of $I_{a,b,c}^{(\delta)}$.

1) $\delta > 2, a = b, c = \infty$.

The identity then reduces to:

$$\frac{1}{\delta-2} \ell_\infty^b(R_{\delta+2}) \overset{\text{(law)}}{=} \frac{1}{\delta} \ell_\infty^b(R_\delta)$$

In fact, both variables are exponentially distributed, with parameters which match the identity; moreover, this identity expresses precisely how the total local time at b for R_δ explodes as $\delta \downarrow 2$.

2) $\delta > 2, a = 0, c = \infty$.

The identity then becomes:

$$\int\limits_0^\infty ds\, 1_{(R_{\delta+2}(s) \le b)} + \frac{b}{\delta-2} \ell_\infty^b(R_{\delta+2}) \overset{\text{(law)}}{=} \int\limits_0^\infty ds\, 1_{(R_\delta(s) \le b)}$$

Considered together with the original C-T identity (4.1), this gives a functional of $R_{\delta+2}$ which is distributed as $T_b(R_{\delta-2})$.

3) $\delta = 2$.

Taking $a = 0$, we obtain:

$$\int\limits_0^\infty ds\, 1_{(R_4(s) \le b)} + b \log\left(\frac{c}{b}\right) \ell_\infty^b(R_4) \overset{\text{(law)}}{=} \int\limits_0^{T_c} ds\, 1_{(R_2(s) \le b)} \ ,$$

whilst taking $a = b > 0$, we obtain:

$$\log\left(\frac{c}{b}\right) \ell_\infty^b(R_4) \overset{\text{(law)}}{=} \frac{1}{2} \ell_{T_c}^b(R_2) \ .$$

In particular, we deduce from these identities in law the following limit results:

$$\frac{1}{\log c} \int_0^{T_c} ds\ 1_{(R_2(s)\le b)} \xrightarrow[c\to\infty]{\text{(law)}} b\ell_\infty^b(R_4)\ ,$$

and

$$\frac{1}{\log c}\ell_{T_c}^b(R_2) \xrightarrow[c\to\infty]{\text{(law)}} 2\ell_\infty^b(R_4)\ .$$

In fact, these limits in law may be seen as particular cases of the Kallianpur-Robbins asymptotic result for additive functionals of planar Brownian motion $(Z_t, t \ge 0)$, which states that:

i) If f belongs to $L^1(\mathbb{C}, dx\ dy)$, and is locally bounded, and if

$$A_t^f \overset{\text{def}}{=} \int_0^t ds\ f(Z_s)\ ,$$

then:

$$\frac{1}{\log t} A_t^f \xrightarrow[t\to\infty]{\text{(law)}} \left(\frac{1}{2\pi}\bar{f}\right) \mathbf{e}\ ,$$

where $\bar{f} = \int_{\mathbb{C}} dx\ dy\ f(x,y)$, and $\mathbf{e}$ is a standard exponential variable.

Moreover, one has:

$$\frac{1}{\log t}\left(A_t^f - A_{T_{\sqrt{t}}}^f\right) \xrightarrow[t\to\infty]{(P)} 0$$

ii) (Ergodic theorem)   If f and g both satisfy the conditions stated in (i), then:

$$A_t^f/A_t^g \xrightarrow[t\to\infty]{\text{a.s}} \bar{f}/\bar{g}\ .$$

4) $\delta < 2, a = 0$.
   The identity in law then becomes:

$$\int_0^\infty ds\ 1_{(R_{\delta+2}(s)\le b)} + b^{\delta-1}\left(\frac{c^{2-\delta} - b^{2-\delta}}{2-\delta}\right) \ell_\infty^b(R_{\delta+2}) \overset{\text{(law)}}{=} \int_0^{T_c} ds\ 1_{(R_\delta(s)\le b)}$$

which, as a consequence, implies:

$$\frac{1}{c^{2-\delta}} \int_0^{T_c} ds \, 1_{(R_\delta(s) \leq b)} \xrightarrow[c \to \infty]{(\text{law})} \frac{b^{\delta-1}}{2-\delta} \ell_\infty^b(R_{\delta+2}) \ . \tag{4.6}$$

In fact, this limit in law can be explained much easier than the limit in law in the previous example. Here is such an explanation:

the local times $(\ell_t^a(R_\delta); a > 0, t \geq 0)$, which, until now in this Chapter, have been associated to the Bessel process R_δ, are the *semimartingale local times,* i.e.: they may be defined via the occupation density formula, with respect to Lebesgue measure on $\mathbb{R}_+$. However, at this point, it is more convenient to define the family $\{\lambda_t^x(R_\delta)\}$ of *diffusion local times* by the formula:

$$\int_0^t ds \, f\left(R_\delta(s)\right) = \int_0^\infty dx \, x^{\delta-1} \lambda_t^x(R_\delta) f(x) \tag{4.7}$$

for every Borel function $f : \mathbb{R}_+ \to \mathbb{R}_+$.

(The advantage of this definition is that the diffusion local time $(\lambda_t^0(R_\delta); t > 0)$ will be finite and strictly positive). Now, we consider the left-hand side of (4.6); we have:

$$\frac{1}{c^2} \int_0^{T_c} ds \, 1_{(R_\delta(s) \leq b)} \stackrel{(\text{law})}{=} \int_0^{T_1} du \, 1_{\left(R_\delta(u) \leq \frac{b}{c}\right)} \qquad \text{(by scaling)}$$

$$\stackrel{(\text{law})}{=} \int_0^{b/c} dx \, x^{\delta-1} \lambda_{T_1}^x(R_\delta) \qquad \text{(by formula (4.7))}$$

$$\stackrel{(\text{law})}{=} \int_0^b \frac{dy}{c} \left(\frac{y}{c}\right)^{\delta-1} \lambda_{T_1}^{y/c}(R_\delta) \quad \text{(by change of variables)}$$

Hence, we have:

$$\frac{1}{c^{2-\delta}} \int_0^{T_c} ds \, 1_{(R_\delta(s) \leq b)} \xrightarrow[c \to \infty]{(\text{law})} \lambda_{T_1}^0(R_\delta) \frac{b^\delta}{\delta} \tag{4.8}$$

The convergence results (4.6) and (4.8) imply:

$$\lambda_{T_1}^0(R_\delta) \frac{b^\delta}{\delta} \stackrel{(\text{law})}{=} \frac{b^{\delta-1}}{2-\delta} \ell_\infty^b(R_{\delta+2}) \tag{4.9}$$

It is not hard to convince oneself directly that the identity in law (4.9) holds; indeed, from the scaling property of $R_{\delta+2}$, we deduce that:

$$\ell_\infty^b(R_{\delta+2}) \stackrel{(\text{law})}{=} b\ell_\infty^1(R_{\delta+2}) \ ,$$

so that formula (4.9) reduces to:

$$\lambda_{T_1}^0(R_\delta) \overset{(\text{law})}{=} \frac{\delta}{2-\delta} \ell_\infty^1(R_{\delta+2}) \tag{4.10}$$

Exercise 4.1 Give a proof of the identity in law (4.10) as a consequence of the Ray-Knight theorems for $\left(\lambda_{T_1}^x(R_\delta); x \leq 1\right)$ and $(\ell_\infty^x(R_{\delta+2}); x \geq 0)$.

Exercise 4.2 Let $c > 0$ be fixed. Prove that: $\dfrac{1}{c-a}\ell_{T_c}^a(R_\delta)$ converges in law, as $a \uparrow c$, and identify the limit in law.

Hint: Either use the identity in law $I_{a,c,c}^{(\delta)}$ or a Ray-Knight theorem for $(\ell_{T_c}^a(R_\delta); a \leq c)$.

4.4 On a computation of Földes-Révész

Földes-Révész [42] have obtained, as a consequence of formulae in Borodin [21] concerning computations of laws of Brownian local times, the following identity in law, for $r > q$:

$$\int_0^\infty dy \, 1_{(0 < \ell_{\tau_r}^y < q)} \overset{(\text{law})}{=} T_{\sqrt{q}}(R_2) \tag{4.11}$$

where, on the left-hand side, $\ell_{\tau_r}^y$ denotes the local time of Brownian motion taken at level y, and at time τ_r, the first time local time at 0 reaches r, and, on the right-hand side, $T_{\sqrt{q}}(R_2)$ denotes the first hitting time of $\sqrt{q}$ by R_2, a two-dimensional Bessel process starting from 0.

We now give an explanation of formula (4.11), using jointly the Ray-Knight theorem and the Ciesielski-Taylor identity in law (4.1).

From the Ray-Knight theorem on Brownian local times up to time τ_r, we know that the left-hand side of formula (4.11) is equal, in law, to:

$$\int_0^{T_0} dy \, 1_{(Y_y < q)} \; ,$$

where $(Y_y, y \geq 0)$ is a BESQ process, with dimension 0, starting from r. Since $r > q$, we may as well assume, using the strong Markov property that $Y_0 = q$, which explains why the law of the left-hand side of (4.11) does not depend on $r(\geq q)$.

Now, by time reversal, we have:

$$\int_0^{T_0} dy \, 1_{(Y_y < q)} \overset{(law)}{=} \int_0^{\mathcal{L}_q} dy \, 1_{(\hat{Y}_y < q)} \ ,$$

where $(\hat{Y}_y, y \geq 0)$ is a BESQ process, with dimension 4, starting from 0, and $\mathcal{L}_q = \sup\{y : \hat{Y}_y = q\}$. Moreover, we have, obviously:

$$\int_0^{\mathcal{L}_q} dy \, 1_{(\hat{Y}_y < q)} = \int_0^{\infty} dy \, 1_{(\hat{Y}_y < q)} \tag{4.12}$$

and we deduce from the original Ciesielski-Taylor identity in law (4.1), taken for $\delta = 2$, together with the scaling property of a BES process starting from 0, that the right-hand side of (4.12) is equal in law to $T_{\sqrt{q}}(R_2)$.

Comments on Chapter 4

The proof, presented here, of the Ciesielski-Taylor identities follows Yor [101]; it combines the RK theorems with the integration by parts formula (2.14). More generally, Biane's extensions of the CT identities to a large class of diffusions ([12]) may also be obtained in the same way. It would be interesting to know whether another family of extensions of the CT identities, obtained by Carmona-Petit-Yor [25] for certain càdlàg Markov processes which are related to Bessel processes through some intertwining relationship, could also be derived from some adequate version of the integration by parts formula (2.14).

In paragraph 4.3, it seemed an amusing exercise to look at some particular cases of the identity in law $I_{a,b,c}^{(\delta)}$, and to relate these examples to some better known relations, possibly of an asymptotic kind.

Finally, paragraph 4.4 presents an interesting application of the CT identities.

Chapter 5

On the winding number of planar BM

The appearance in Chapter 3 of Bessel processes of various dimensions is very remarkable, despite the several proofs of the Ray-Knight theorems on Brownian local times which have now been published.

It is certainly less astonishing to see that the 2-dimensional Bessel process plays an important part in the study of the windings of planar Brownian motion $(Z_t, t \geq 0)$; indeed, one feels that, when Z wanders far away from 0, or, on the contrary, when it gets close to 0, then it has a tendency to wind more than when it lies in the annulus, say $\{z : r \leq |z| \leq R\}$, for some given $0 < r < R < \infty$. However, some other remarkable feature occurs: the computation of the law, for a fixed time t, of the winding number θ_t of $(Z_u, u \leq t)$ around 0, is closely related to the knowledge of the semigroups of all Bessel processes, with dimensions δ varying between 2 and ∞.

There have been, in the 1980's, a number of studies about the asymptotics of winding numbers of planar BM around a finite set of points (see, e.g., a short summary in [81], Chapter XII). It then seemed more interesting to develop here some exact computations for the law of the winding number up to a fixed time t, for which some open questions still remain.

5.1 Preliminaries

(5.1.1) Consider $Z_t = X_t + iY_t, t \geq 0$, a planar BM starting from $z_0 \neq 0$. We have

Proposition 5.1 *1) With probability 1, $(Z_t, t \geq 0)$ does not visit 0.*

*2) A continuous determination of the logarithm along the trajectory $(Z_u(\omega),$
$u \geq 0)$ is given by the stochastic integral:*

$$\log_\omega (Z_t(\omega)) - \log_\omega (Z_0(\omega)) = \int_0^t \frac{dZ_u}{Z_u} \, , \qquad t \geq 0 \, . \tag{5.1}$$

We postpone the proof of Proposition 5.1 for a moment, in order to write
down the following pair of formulae, which are immediate consequences
of (5.1):

$$\log |Z_t(\omega)| - \log |Z_0(\omega)| = Re \int_0^t \frac{dZ_u}{Z_u} = \int_0^t \frac{X_u dX_u + Y_u dY_u}{|Z_u|^2} \tag{5.2}$$

and

$$\theta_t(\omega) - \theta_0(\omega) = Im \int_0^t \frac{dZ_u}{Z_u} = \int_0^t \frac{X_u dY_u - Y_u dX_u}{|Z_u|^2} \, , \tag{5.3}$$

where $(\theta_t(\omega), t \geq 0)$ denotes a continuous determination of the argument of
$(Z_u(\omega), u \leq t)$ around 0.

We now note that:

$$\beta_u = \int_0^u \frac{X_s dX_s + Y_s dY_s}{|Z_s|} \quad (u \geq 0) \quad \text{and} \quad \gamma_u = \int_0^u \frac{X_s dY_s - Y_s dX_s}{|Z_s|} \quad (u \geq 0)$$

are two orthogonal martingales with increasing processes $\langle \beta \rangle_u = \langle \gamma \rangle_u \equiv$
u, hence, they are two *independent* Brownian motions; moreover, it is not
difficult to show that:

$$\mathcal{R}_t \stackrel{\text{def}}{=} \sigma \{|Z_u|, u \leq t\} \equiv \sigma \{\beta_u, u \leq t\} \, , \quad \text{up to negligible sets;}$$

hence, we have the following:

for $\nu \in \mathbb{R}$, $E\left[\exp\left(i\nu(\theta_t - \theta_0)\right) \mid \mathcal{R}_\infty\right] = \exp\left(-\frac{\nu^2}{2} H_t\right)$, where $H_t \stackrel{\text{def}}{=} \int_0^t \frac{ds}{|Z_s|^2}$

$$\tag{5.4}$$

This formula shall be of great help, in the next paragraph, to compute the
law of θ_t, for fixed t.

(5.1.2) We now prove Proposition 5.1. The first statement of Proposition 5.1 follows from

Proposition 5.2 *(B. Davis [28]) If $f : \mathbb{C} \to \mathbb{C}$ is holomorphic and not constant, then there exists a planar BM $(\hat{Z}_t, t \geq 0)$ such that:*

$$f(Z_t) = \hat{Z}_{A_t^f} \ , \quad t \geq 0 \ , \quad \text{and} \quad A_\infty^f = \infty \quad a.s.$$

To prove Proposition 5.1, we apply Proposition 5.2 with $f(z) = \exp(z)$. Then,

$$\exp(Z_t) = \hat{Z}_{A_t}, \text{ with: } A_t = \int_0^t ds \exp(2X_s) \ .$$

The "trick" is to consider, instead of Z, the planar BM $(\hat{Z}_u, u \geq 0)$, which starts from $\exp(Z_0) \neq 0$, at time $t = 0$, and shall never reach 0, since $\exp(z) \neq 0$, for every $z \in \mathbb{C}$.

Next, to prove formula (5.1), it suffices to show:

$$\exp \left(\int_0^t \frac{dZ_u}{Z_u} \right) = \frac{Z_t}{Z_0} \ , \quad \text{for all } t \geq 0 \ . \tag{5.5}$$

This follows from Itô's formula, from which we easily deduce:

$$d \left(\frac{1}{Z_t} \exp \left(\int_0^t \frac{dZ_u}{Z_u} \right) \right) = 0$$

Exercise 5.1 Give another proof of the identity (5.5), using the uniqueness of solutions of the stochastic equation:

$$U_t = Z_0 + \int_0^t U_s \frac{dZ_s}{Z_s} \ , \quad t \geq 0 \ .$$

(5.1.3) In the sequel, we shall also need the two following formulae, which involve the modified Bessel functions I_ν.

a) The semi-group $P_t^{(\nu)}(r, d\rho) = p_t^{(\nu)}(r, \rho) d\rho$ of the Bessel process of index $\nu > 0$ is given by the formula:

$$p_t^{(\nu)}(r,\rho) = \frac{1}{t}\left(\frac{\rho}{r}\right)^{\nu}\rho\exp-\left(\frac{r^2+\rho^2}{2t}\right)I_\nu\left(\frac{r\rho}{t}\right) \qquad (r,\rho,t>0)$$

(see, e.g., Revuz-Yor [81], p. 411).

b) For any $\lambda \in \mathbb{R}$, and $r > 0$, the modified Bessel function $I_\lambda(r)$ admits the following integral representation:

$$I_\lambda(r) = \frac{1}{\pi}\int\limits_0^\pi d\theta(\exp(r\cos\theta))\cos(\lambda\theta) - \frac{\sin(\lambda\pi)}{\pi}\int\limits_0^\infty du\, e^{-rchu-\lambda u}$$

(see, e.g., Watson [90], and Lebedev [63], formula (5.10.8), p. 115).

5.2 Explicit computation of the winding number of planar Brownian motion

(5.2.1) With the help of the preliminaries, we shall now prove the following

Theorem 5.1 *For any $z_0 \neq 0$, $r, t > 0$, and $\nu \in \mathbb{R}$, we have:*

$$E_{z_0}\left[\exp\left(i\nu(\theta_t - \theta_0)\right)\Big|\,|Z_t| = \rho\right] = \frac{I_{|\nu|}}{I_0}\left(\frac{|z_0|\rho}{t}\right) \qquad (5.6)$$

Before we prove formula (5.6), let us comment that this formula shows in particular that, for every given $r > 0$, the function: $\nu \to \frac{I_{|\nu|}}{I_0}(r)$ is the Fourier transform of a probability measure, which we shall denote by μ_r; this distribution was discovered, by analytic means, by Hartman-Watson [48], hence we shall call μ_r the Hartman-Watson distribution with parameter r. Hence, μ_r is characterized by:

$$\frac{I_{|\nu|}}{I_0}(r) = \int\limits_{-\infty}^\infty \exp(i\nu\theta)\mu_r(d\theta) \qquad (\nu \in \mathbb{R})\ . \qquad (5.7)$$

The proof of Theorem 5.1 shall follow from

Proposition 5.3 *Let* $r > 0$. *For any* $\nu \geq 0$, *define* $P_r^{(\nu)}$ *to be the law of the Bessel process, with index* ν, *starting at* r, *on the canonical space* $\Omega_+^* \equiv C(\mathbb{R}_+, \mathbb{R}_+)$.

Then, we have:

$$P_r^{(\nu)}\Big|_{\mathcal{R}_t} = \left(\frac{R_t}{r}\right)^\nu \exp\left(-\frac{\nu^2}{2}H_t\right) \cdot P_r^{(0)}\Big|_{\mathcal{R}_t} . \tag{5.8}$$

PROOF: This is a simple consequence of Girsanov's theorem.
However, remark that the relation (5.8) may also be considered as a variant of the simpler Cameron-Martin relation:

$$W^{(\nu)}\Big|_{\mathcal{F}_t} = \exp\left(\nu X_t - \frac{\nu^2 t}{2}\right) \cdot W\Big|_{\mathcal{F}_t} \tag{5.9}$$

where $W^{(\nu)}$ denotes the law, on $C(\mathbb{R}_+, \mathbb{R})$, of Brownian motion with drift ν. Formula (5.9) implies (5.8) after time-changing, since, under $P_r^{(\nu)}$, one has:

$$R_t = r\exp(B_u + \nu u)\Big|_{u = H_t} , \quad \text{and} \quad H_t = \inf\left\{u : \int_0^u ds\, \exp 2(B_s + \nu s) > t\right\}$$

$\square$

We now finish the proof of Theorem 5.1; from formulae (5.2) and (5.3), and the independence of β and γ, we deduce, denoting $r = |z_0|$, that:

$$E_{z_0}\left[\exp\left(i\nu(\theta_t - \theta_0)\right)\Big| \, |Z_t| = \rho\right] = E_{z_0}\left[\exp\left(-\frac{\nu^2}{2}H_t\right)\Big| \, |Z_t| = \rho\right]$$

$$= E_r^{(0)}\left[\exp\left(-\frac{\nu^2}{2}H_t\right)\Big| R_t = \rho\right]$$

Now, from formula (5.8), we deduce that for every Borel function $f : \mathbb{R}_+ \to \mathbb{R}_+$, we have:

$$E_r^{(\nu)}[f(R_t)] = E_r^{(0)}\left[f(R_t)\left(\frac{R_t}{t}\right)^\nu \exp\left(-\frac{\nu^2}{2}H_t\right)\right] ,$$

which implies:

$$p_t^{(\nu)}(r, \rho) = p_t^{(0)}(r, \rho)\left(\frac{\rho}{r}\right)^\nu E_r^{(0)}\left[\exp\left(-\frac{\nu^2}{2}H_t\right)\Big| R_t = \rho\right] ,$$

and formula (5.6) now follows immediately from the explicit expressions of $p_t^{(\nu)}(r,\rho)$ and $p_t^{(0)}(r,\rho)$ given in (5.1.3). □

With the help of the classical integral representation of I_λ, which was presented above in (5.1.3), we are able to give the following explicit additive decomposition of μ_r.

Theorem 5.2 *1) For any $r > 0$, we have*

$$\mu_r(d\theta) = p_r(d\theta) + q_r(d\theta) \ ,$$

where: $p_r(d\theta) = \frac{1}{2\pi I_0(r)} \exp(r\cos\theta)1_{[-\pi,\pi[}(\theta)d\theta$ is the Von Mises distribution with parameter r, and $q_r(d\theta)$ is a bounded signed measure, with total mass equal to 0.

2) q_r admits the following representation:

$$q_r(d\theta) = \frac{1}{I_0(r)} \left\{ -e^{-r}m + m * \int_0^\infty \pi_r(du)c_u \right\} \ ,$$

where:

$$m(d\theta) = \frac{1}{2\pi}1_{[-\pi,\pi[}(\theta)d\theta; \ \pi_r(du) = e^{-r\,chu}r(shu)du; \ c_u(d\theta) = \frac{ud\theta}{\pi(\theta^2 + u^2)}$$

3) q_r may also be written as follows:

$$q_r(d\theta) = \frac{1}{2\pi I_0(r)} \{\Phi_r(\theta - \pi) - \Phi_r(\theta + \pi)\} d\theta \ ,$$

where

$$\Phi_r(x) = \int_0^\infty dt \ e^{-r\,cht}\frac{x}{\pi(t^2 + x^2)} = \int_0^\infty \pi_r(dt)\frac{1}{\pi}\text{Arc tg}\left(\frac{t}{x}\right) \ . \qquad (5.10)$$

It is a tantalizing question to interpret precisely every ingredient in the above decomposition of μ_r in terms of the winding number of planar Brownian motion. This is simply solved for p_r, which is the law of the principal determination α_t (e.g.: with values in $[-\pi, \pi[$), of the argument of the random variable Z_t, given $R_t \equiv |Z_t|$, a question which does not involve the complicated manner in which Brownian motion $(Z_u, u \leq t)$ has wound around 0 up to time t, but depends only on the distribution of the 2-dimensional random variable Z_t.

On the contrary, the quantities which appear in the decomposition of q_r in the second statement of Theorem 5.2 are not so easy to interpret. However, the Cauchy distribution c_1 which appears there is closely related to Spitzer's asymptotic result, which we now recall.

Theorem 5.3 As $t \to \infty$, $\dfrac{2\theta_t}{\log t} \xrightarrow[t\to\infty]{\text{(law)}} C_1$, where C_1 is a Cauchy variable with parameter 1.

PROOF: Following Itô-McKean ([50], p. 270) we remark that, from the convergence in law of $\dfrac{R_t}{\sqrt{t}}$, as $t \to \infty$, it is sufficient, to prove the theorem, to show that, for every $\nu \in \mathbb{R}$, we have:

$$E_{z_0}\left[\exp\left(\frac{2i\nu\theta_t}{\log t}\right) \mid R_t = \rho\sqrt{t}\right] \xrightarrow[t\to\infty]{} \exp\left(-|\nu|\right) ,$$

which, thanks to formula (5.6), amounts to showing:

$$\frac{I_\lambda}{I_0}\left(\frac{|z_0|\rho}{\sqrt{t}}\right) \xrightarrow[t\to\infty]{} e^{-|\nu|} \tag{5.11}$$

with the notation: $\lambda = \dfrac{2|\nu|}{\log t}$.

Making an integration by parts in the integral representation of $I_\lambda(r)$ in (5.1.3), b), the proof of (5.11) shall be finished once we know that:

$$\int \pi_{p/\sqrt{t}}(du) \exp\left(-\frac{|\nu|u}{\log\sqrt{t}}\right) \xrightarrow[t\to\infty]{} e^{-|\nu|} ,$$

where $p = \rho|z_0|$. However, if we consider the linear application $\ell_t : u \to \frac{u}{\log t}$ ($u \in \mathbb{R}_+$), it is easily shown that: $\ell_t(\pi_{p/t}) \xrightarrow[t\to\infty]{(w)} \varepsilon_1(du)$, i.e.: the image of $\pi_{p/t}(du)$ by ℓ_t converges weakly, as $t \to \infty$, to the Dirac measure at 1. $\square$

The finite measure $\pi_r(du)$ appears also naturally in the following representation of the law of the winding number around 0 of the "Brownian lace" (= complex Brownian bridge) with extremity $z_0 \neq 0$, and length t.

Theorem 5.4 *Let* $W = \dfrac{\theta_t}{2\pi}$ *be the winding number of the Brownian lace of length* t, *starting and ending at* z_0. *Then, with the notation:* $r = \dfrac{|z_0|^2}{t}$, *we have:*

$$W \overset{\text{(law)}}{=} \varepsilon \left[\frac{C_T}{2\pi} + \frac{1}{2} \right] \tag{5.12}$$

where T *is a random variable with values in* $\mathbb{R}_+$, *such that:*

$$P(T \in du) = e^r \pi_r(du) \ ,$$

ε *takes the values 0 and 1, with probabilities:*

$$P(\varepsilon = 0) = 1 - e^{-2r} \ , \qquad P(\varepsilon = 1) = e^{-2r} \ ,$$

$(C_u)_{u \geq 0}$ *is a symmetric Cauchy process starting from 0,* T, ε *and* $(C_u)_{u \geq 0}$ *are independent, and, finally,* $[x]$ *denotes the integer part of* $x \in \mathbb{R}$.

For the sake of clarity, we shall now assume, while proving Theorem 5.4, that $z_0 = |z_0|$; there is no loss of generality, thanks to the conformal invariance of Brownian motion.

In particular, we may choose $\theta_0 = 0$, and it is then easy to deduce from the identity (5.6), and the representation of μ_r given in Theorem 5.2 that, for any Borel function
$f : \mathbb{R} \to \mathbb{R}_+$, one has:

$$E_{z_0}\left[f(\theta_t) \mid Z_t = z\right] = f(\alpha_t) + e^{-\tilde{r}\cos(\alpha_t)} \sum_{n \in \mathbb{Z}} a_n(t, \tilde{r}) f(\alpha_t + 2n\pi) \qquad (*)$$

where α_t is equal, as above, to the determination of the argument of the variable Z_t in $]-\pi, \pi]$, $\tilde{r} = \dfrac{|z_0| \, |z|}{t}$, and

$$a_n(\tilde{r}, t) = \Phi_{\tilde{r}}(\alpha_t + (2n-1)\pi) - \Phi_{\tilde{r}}(\alpha_t + (2n+1)\pi) \ .$$

In particular, for $z = z_0$, one has: $r = \tilde{r}$, $\alpha_t = 0$, and the previous formula $(*)$ becomes, for $n \neq 0$:

$$P_{z_0}\left(\theta_t = 2n\pi \mid Z_t = z_0\right)$$

$$= \int_0^\infty \pi_r(du)e^{-r}\frac{1}{\pi}\left\{\text{Arc tg}\left(\frac{u}{(2n-1)\pi}\right) - \text{Arc tg}\left(\frac{u}{(2n+1)\pi}\right)\right\} \text{ (from (5.10))}$$

$$= \int_0^\infty \pi_r(du)e^{-r}\int_{(2n-1)\pi}^{(2n+1)\pi}\frac{u\,dx}{\pi(u^2+x^2)}$$

$$= \int P(T \in du)e^{-2r}P\left((2n-1)\pi \leq C_u \leq (2n+1)\pi\right) \qquad (5.13)$$

Likewise, for $n = 0$, one deduces from $(*)$ that:

$$P_{z_0}\left(\theta_t = 0 \mid Z_t = z_0\right) = \int P(T \in du)\left\{(1-e^{-2r}) + e^{-2r}P(-\pi \leq C_u \leq \pi)\right\}$$

$$(5.14)$$

The representation (5.12) now follows from the two formulae (5.13) and (5.14). From Theorem 5.4, we deduce the following interesting

Corollary 5.4.1 *Let θ_t^* be the value at time t of a continuous determination of the argument of the Brownian lace $(Z_u, u \leq t)$, such that $Z_0 = Z_t = z_0$. Then, one has:*

$$\frac{1}{\log t}\theta_t^* \xrightarrow[t\to\infty]{\text{(law)}} C_1 \qquad (5.15)$$

Remark: Note that, in contrast with the statement in Theorem 5.3, the asymptotic winding θ_t^* of the "long" Brownian lace $(Z_u, u \leq t)$, as $t \to \infty$, may be thought of as the sum of the windings of two independent "free" Brownian motions considered on the interval $[0, t]$; it is indeed possible to justify directly this assertion. $\square$

PROOF OF THE COROLLARY: From the representation (5.12), it suffices to show that:

$$\frac{1}{\log t}C_T \xrightarrow[t\to\infty]{\text{(law)}} C_1 \;,$$

where T is distributed as indicated in Theorem 5.4.

Now, this convergence in law follows from

$$C_T \overset{\text{(law)}}{=} T\,C_1\;,$$

and the fact, already seen at the end of the proof of Theorem 5.3, that:

$$\frac{T}{\log t} \xrightarrow[t\to\infty]{(P)} 1 \ .$$

□

(5.2.2)   In order to understand better the representation of W given by formula (5.12), we shall now replace the Brownian lace by a planar Brownian motion with drift, thanks to the invariance of Brownian motion by time-inversion. From this invariance property, we first deduce the following easy

Lemma 5.1 *Let $z_1, z_2 \in \mathbb{C}$, and let $P_{z_1}^{z_2}$ be the law of $(z_1 + \hat{Z}_u + uz_2; u \geq 0)$, where $(\hat{Z}_u, u \geq 0)$ is a planar BM starting from 0.*

Then, the law of $\left(uZ\left(\frac{1}{u}\right), u > 0\right)$ under $P_{z_1}^{z_2}$ is $P_{z_2}^{z_1}$.

As a consequence, we obtain: for every positive functional F,

$$E_{z_0}\left[F(Z_u, u \leq t) \mid Z_t = z\right] = E_{z/t}^{z_0}\left[F\left(uZ\left(\frac{1}{u} - \frac{1}{t}\right); u \leq t\right)\right].$$

We may now state the following

Theorem 5.5 *Let $Z_u = X_u + iY_u$, $u \geq 0$, be a $\mathbb{C}$-valued process, and define $T_t = \inf\{u \leq t : X_u = 0\}$, with $T_t = t$, if $\{\,\}$ is empty, and $\mathcal{L} = \sup\{u : X_u = 0\}$, with $\mathcal{L} = 0$ if $\{\,\}$ is empty.*
Then, we have:

1) for any Borel function $f : \mathbb{R} \times \mathbb{R}_+ \to \mathbb{R}_+$,

$$E_{z_0}\left[f\left(\theta_t, \frac{1}{t} - \frac{1}{T_t}\right) \mid Z_t = z\right] = E_{z/t}^{z_0}\left[f(\theta_\infty, \mathcal{L})\right] \ ;$$

2) moreover, when we take $z_0 = z$, we obtain, with the notation of Theorem 5.4:

$$E_{z_0}\left[f(\theta_t)1_{(T_t < t)} \mid Z_t = z_0\right] = E_{z_0/t}^{z_0}\left[f(\theta_\infty)1_{(\mathcal{L} > 0)}\right]$$
$$= E\left[f\left(2\pi\varepsilon\left[\frac{C_T}{2\pi} + \frac{1}{2}\right]\right)\varepsilon\right]$$

The proof of Theorem 5.5 follows easily from Lemma 5.1 and Theorem 5.4.

Comments on Chapter 5

The computations presented in paragraph 5.1 are, by now, well-known; the development in paragraph 5.2 is taken partly from Yor [98]; some related computations are found in Berger-Roberts [5].

It is very interesting to compare the proof of Theorem 5.3, which follows partly Itô - Mc Kean ([50] and, in fact, the original proof of Spitzer [85]) and makes use of some asymptotics of the modified Bessel functions, with the "computation-free" arguments of Williams (1974; unpublished) and Durrett [37] discussed in detail in Messulam-Yor [65] and Pitman-Yor [75].

It would be interesting to obtain a better understanding of the identity in law (5.12), an attempt at which is presented in Theorem 5.5.

Chapter 6

On some exponential functionals of Brownian motion and the problem of Asian options

In the asymptotic study of the winding number of planar BM made in the second part of Chapter 5, we saw the important role played by the representation of $(R_t, t \geq 0)$, the 2-dimensional Bessel process, as:

$$R_t = \exp(B_{H_t}) , \quad \text{where } H_t = \int_0^t \frac{ds}{R_s^2} ,$$

with $(B_u, u \geq 0)$ a real-valued Brownian motion.

In this chapter, we are interested in the law of the exponential functional:

$$\int_0^t ds \exp(aB_s + bs) ,$$

where $a, b \in \mathbb{R}$, and $(B_s, s \geq 0)$ is a 1-dimensional Brownian motion. To compute this distribution, we can proceed in a manner which is similar to that used in the second part of Chapter 5, in that we also rely upon the exact knowledge of the semigroups of the Bessel processes.

The problem which motivated the development in this chapter is that of the so-called Asian options which, on the mathematical side, consists in computing as explicitly as possible the quantity:

$$C^{(\nu)}(t, k) = E\left[(A_t^{(\nu)} - k)^+\right] , \tag{6.1}$$

where $k, t \geq 0$, and:

$$A_t^{(\nu)} = \int\limits_0^t ds \exp 2(B_s + \nu s) \ ,$$

with B a real-valued Brownian motion starting from 0.

The method alluded to above, and developed in detail in [102], yields an explicit formula for the law of $A_t^{(\nu)}$, and even for that of the pair $(A_t^{(\nu)}, B_t)$. However, then, the density of this law is given in an integral form, and it seems difficult to use this result to obtain a "workable" formula for (6.1).

It is, in fact, easier to consider the Laplace transform in t of $C^{(\nu)}(t, k)$, that is:

$$\lambda \int\limits_0^\infty dt \ e^{-\lambda t} E\left[(A_t^{(\nu)} - k)^+\right] \equiv E\left[(A_{T_\lambda}^{(\nu)} - k)^+\right] \ ,$$

where T_λ denotes an exponential variable with parameter λ, which is independent of B. It is no more difficult to obtain a closed form formula for $E\left[\left\{(A_{T_\lambda}^{(\nu)} - k)^+\right\}^n\right]$ for any $n \geq 0$, and, therefore, we shall present the main result of this chapter in the following form.

Theorem 6.1 *Consider $n \geq 0$ (n is not necessarily an integer) and $\lambda > 0$. Define $\mu = \sqrt{2\lambda + \nu^2}$. We assume that: $\lambda > 2n(n + \nu)$, which is equivalent to: $\mu > \nu + 2n$. Then, we have, for every $x > 0$:*

$$E\left[\left\{\left(A_{T_\lambda}^{(\nu)} - \frac{1}{2x}\right)^+\right\}^n\right] = \frac{E\left[(A_{T_\lambda}^{(\nu)})^n\right]}{\Gamma\left(\frac{\mu-\nu}{2} - n\right)} \int\limits_0^x dt \ e^{-t} \ t^{\frac{\mu-\nu}{2} - n - 1} \left(1 - \frac{t}{x}\right)^{\frac{\mu+\nu}{2} + n}$$

$$(6.2)$$

Moreover, we have:

$$E\left[(A_{T_\lambda}^{(\nu)})^n\right] = \frac{\Gamma(n+1)\Gamma\left(\frac{\mu+\nu}{2} + 1\right)\Gamma\left(\frac{\mu-\nu}{2} - n\right)}{2^n \Gamma\left(\frac{\mu-\nu}{2}\right)\Gamma\left(n + \frac{\mu+\nu}{2} + 1\right)} \ . \qquad (6.3)$$

In the particular case where n is an integer, this formula simplifies into:

$$E\left[(A_{T_\lambda}^{(\nu)})^n\right] = \frac{n!}{\prod\limits_{j=1}^n (\lambda - 2(j^2 + j\nu))} \ . \qquad (6.4)$$

Remarks:

1) It is easily verified, using dominated convergence, that, as $x \to \infty$, both sides of (6.2) converge towards $E\left[(A_{T_\lambda}^{(\nu)})^n\right]$.

2) It appears clearly from formula (6.2) that, in some sense, a first step in the computation of the left-hand side of this formula is the computation of the moments of $A_{T_\lambda}^{(\nu)}$. In fact, in paragraph (6.1), we shall first show how to obtain formula (6.4), independently from the method used in the sequel of the chapter.

6.1 The integral moments of $A_t^{(\nu)}$

In order to simplify the presentation, and to extend easily some of the computations made in the Brownian case to some other processes with independent increments, we shall write, for $\lambda \in \mathbb{R}$

$$E\left[\exp(\lambda B_t)\right] = \exp\left(t\varphi(\lambda)\right) \quad, \quad \text{where, here,} \quad \varphi(\lambda) = \frac{\lambda^2}{2} \ . \tag{6.5}$$

We then have the following

Theorem 6.2

1) Let $\mu \geq 0$, $n \in \mathbb{N}$, and $\alpha > \varphi(\mu + n)$. Then, the formula:

$$\int_0^\infty dt \exp(-\alpha t) E\left[\left(\int_0^t ds \exp(B_s)\right)^n \exp(\mu B_t)\right] = \frac{n!}{\displaystyle\prod_{j=0}^n (\alpha - \varphi(\mu + j))} \tag{6.6}$$

holds.

2) Let $\mu \geq 0$, $n \in \mathbb{N}$, and $t \geq 0$. Then, we have:

$$E\left[\left(\int_0^t ds \exp B_s\right)^n \exp(\mu B_t)\right] = E\left[P_n^{(\mu)}(\exp B_t) \exp(\mu B_t)\right] \ , \tag{6.7}$$

where $(P_n^{(\mu)}, n \in \mathbb{N})$ is the following sequence of polynomials:

$$P_n^{(\mu)}(z) = n! \sum_{j=0}^{n} c_j^{(\mu)} z^j \;, \quad \text{with } c_j^{(\mu)} = \prod_{\substack{k \neq j \\ 0 \leq k \leq n}} (\varphi(\mu + j) - \varphi(\mu + k))^{-1} \;.$$

Remark: With the following modifications, this theorem may be applied to a large class of processes with independent increments:

i) we assume that (X_t) is a process with independent increments which admits exponential moments of all orders;
under this only condition, formula (6.6) is valid for α large enough;

ii) Let φ be the Lévy exponent of X which is defined by:

$$E_0\left[\exp(mX_s)\right] = \exp\left(s\varphi(m)\right) \;.$$

Then, formula (6.7) also extends to (X_t), provided $\varphi\big|_{\mathbb{R}_+}$ is injective, which implies that the argument concerning the additive decomposition formula in the proof below still holds.

PROOF OF THEOREM 6.2

1) We define

$$\phi_{n,t}(\mu) = E\left[\left(\int_0^t ds \exp(B_s)\right)^n \exp(\mu B_t)\right]$$

$$= n! E\left[\int_0^t ds_1 \int_0^{s_1} ds_2 \ldots \int_0^{s_{n-1}} ds_n \exp(B_{s_1} + \cdots + B_{s_n} + \mu B_t)\right]$$

We then remark that

$$E\left[\exp(\mu B_t + B_{s_1} + \cdots + B_{s_n})\right]$$
$$= E\left[\exp\left\{\mu(B_t - B_{s_1}) + (\mu + 1)(B_{s_1} - B_{s_2}) + \cdots + (\mu + n)B_{s_n}\right\}\right]$$
$$= \exp\left\{\varphi(\mu)(t - s_1) + \varphi(\mu + 1)(s_1 - s_2) + \cdots + \varphi(\mu + n)s_n\right\} \;.$$

Therefore, we have:

$$\int_0^\infty dt \exp(-\alpha t)\phi_{n,t}(\mu)$$

$$= n!\int_0^\infty dt\, e^{-\alpha t}\int_0^t ds_1\int_0^{s_1} ds_2\int_0^{s_{n-1}} ds_n \exp\{\varphi(\mu)(t-s_1)+\cdots+\varphi(\mu+n)s_n\}$$

$$= n!\int_0^\infty ds_n \exp(-(\alpha-\varphi(\mu+n))s_n)\ldots$$

$$\ldots\int_{s_n}^\infty ds_{n-1}\exp(-(\alpha-\varphi(\mu+n-1))(s_{n-1}-s_n))\int_{s_1}^\infty dt\exp(-(\alpha-\varphi(\mu))(t-s_1)),$$

so that, in the case: $\alpha > \varphi(\mu+n)$, we obtain formula (6.6) by integrating successively the $(n+1)$ exponential functions.

2) Next, we use the additive decomposition formula:

$$\frac{1}{\prod_{j=0}^n (\alpha-\varphi(\mu+j))} = \sum_{j=0}^n c_j^{(\mu)}\frac{1}{(\alpha-\varphi(\mu+j))}$$

where $c_j^{(\mu)}$ is given as stated in the Theorem, and we obtain, for $\alpha > \varphi(\mu+n)$:

$$\int_0^\infty dt\, e^{-\alpha t}\phi_{n,t}(\mu) = n!\sum_{j=0}^n c_j^{(\mu)}\int_0^\infty dt\, e^{-\alpha t}e^{\varphi(\mu+j)t}$$

a formula from which we deduce:

$$\phi_{n,t}(\mu) = n!\sum_{j=0}^n c_j^{(\mu)}\exp(\varphi(\mu+j)t) = n!\sum_{j=0}^n c_j^{(\mu)}E[\exp(jB_t)\exp(\mu B_t)]$$
$$= E\left[P_n^{(\mu)}(\exp B_t)\exp(\mu B_t)\right]\quad.$$

Hence, we have proved formula (6.7). $\square$

As a consequence of Theorem 6.2, we have the following

Corollary 6.2.1 *For any $\lambda \in \mathbb{R}$, and any $n \in \mathbb{N}$, we have:*

$$\lambda^{2n} E\left[\left(\int_0^t du \exp(\lambda B_u)\right)^n\right] = E[P_n(\exp \lambda B_t)] \tag{6.8}$$

where

$$P_n(z) = 2^n(-1)^n \left\{ \frac{1}{n!} + 2\sum_{j=1}^{n} \frac{n!(-z)^j}{(n-j)!(n+j)!} \right\} . \qquad (6.9)$$

PROOF: Thanks to the scaling property of Brownian motion, it suffices to prove formula (6.8) for $\lambda = 1$, and any $t \geq 0$. In this case, we remark that formula (6.8) is then precisely formula (6.7) taken with $\mu = 0$, once the coefficients $c_j^{(0)}$ have been identified as:

$$c_0^{(0)} = (-1)^n \frac{2^n}{(n!)^2} \quad ; \quad c_j^{(0)} = \frac{2^n(-1)^{n-j}2}{(n-j)!(n+j)!} \quad (1 \leq j \leq n) \quad ;$$

therefore, it now appears that the polynomial P_n is precisely $P_n^{(0)}$, and this ends the proof. $\square$

It may also be helpful to write down explicitly the moments of $A_t^{(\nu)}$.

Corollary 6.2.2 *For any $\lambda \in \mathbb{R}^*$, $\mu \in \mathbb{R}$, and $n \in \mathbb{N}$, we have:*

$$\lambda^{2n} E\left[\left(\int_0^t du \exp \lambda(B_u + \mu u) \right)^n \right] = n! \sum_{j=0}^{n} c_j^{(\mu/\lambda)} \exp\left(\left(\frac{\lambda^2 j^2}{2} + \lambda j \mu \right) t \right) . \qquad (6.10)$$

In particular, we have, for $\mu = 0$

$$\lambda^{2n} E\left[\left(\int_0^t du \exp \lambda B_u \right)^n \right] = n! \left\{ \frac{(-1)^n}{(n!)^2} + 2\sum_{j=1}^{n} \frac{(-1)^{n-j}}{(n-j)!(n+j)!} \exp \frac{\lambda^2 j^2 t}{2} \right\} \qquad (6.11)$$

6.2 A study in a general Markovian set-up

It is interesting to give a theoretical solution to the problem of Asian options in a general Markovian set-up, for the two following reasons, at least:

- on one hand, the general presentation allows to understand simply the nature of the quantities which appear in the computations;

- on the other hand, this general approach may allow to choose some other stochastic models than the geometric Brownian motion model.

Therefore, we consider $\{(X_t), (\theta_t), (P_x)_{x \in E}\}$ a strong Markov process, and $(A_t, t \geq 0)$ a continuous additive functional, which is strictly increasing, and such that: $P_x(A_\infty = \infty) = 1$, for every $x \in E$.

Consider, moreover, $g : \mathbb{R} \to \mathbb{R}_+$, a Borel function such that $g(x) = 0$ if $x \leq 0$. (In the applications, we shall take: $g(x) = (x^+)^n$).

Then, define:

$$G_x(t) = E_x[g(A_t)] \quad , \quad G_x(t, k) = E_x[g(A_t - k)]$$

and

$$G_x^{(\lambda)}(k) = E_x\left[\int_0^\infty dt\, e^{-\lambda t} g(A_t - k)\right] .$$

We then have the important

Proposition 6.1 *Define $\tau_k = \inf\{t : A_t \geq k\}$. The two following formulae hold:*

$$G_x^{(\lambda)}(k) = \int_0^\infty dv\, e^{-\lambda v} E_x\left[e^{-\lambda \tau_k} G_{X_{\tau_k}}(v)\right] \tag{6.12}$$

and, if g is increasing, and absolutely continuous,

$$G_x^{(\lambda)}(k) = \frac{1}{\lambda} \int_k^\infty dv\, g'(v - k) E_x[e^{-\lambda \tau_v}] . \tag{6.13}$$

Remark: In the application of these formulae to Brownian motion, we shall see that the equality between the right-hand sides of formulae (6.12) and (6.13) is the translation of a classical "intertwining" identity between confluent hypergeometric functions. This is one of the reasons why it seems important to insist upon this identity; in any case, this discussion shall be taken up in paragraph 6.5.

PROOF OF PROPOSITION 6.1:

1) We first remark that, on the set $\{t \geq \tau_k\}$, the following relation holds:

$$A_t(\omega) = A_{\tau_k}(\omega) + A_{t-\tau_k}(\theta_{\tau_k}\omega) = k + A_{t-\tau_k}(\theta_{\tau_k}\omega) ;$$

then, using the strong Markov property, we obtain:

$$G_x(t,k) \equiv E_x[g(A_t - k)] = E_x\left[E_{X_{\tau_k}(\omega)}\left[g(A_{t-\tau_k(\omega)})\right]1_{(\tau_k(\omega)\le t)}\right] \quad ;$$

hence: $G_x(t,k) = E_x\left[G_{X_{\tau_k}}(t - \tau_k)1_{(\tau_k\le t)}\right].$

This implies, using Fubini's theorem:

$$G_x^{(\lambda)}(k) = E_x\left[\int_{\tau_k}^{\infty} dt\, e^{-\lambda t}G_{X_{\tau_k}}(t - \tau_k)\right] \quad ,$$

and formula (6.12) follows.

2) Making the change of variables $t = v - k$ in the integral in (6.13) and using the strong Markov property, we may write the right-hand side of (6.13) as:

$$\frac{1}{\lambda}E_x\left[\int_0^{\infty} dt\, g'(t)e^{-\lambda\tau_k}E_{X_{\tau_k}}(e^{-\lambda\tau_t})\right]$$

Therefore, in order to prove that the right-hand sides of (6.12) and (6.13) are equal, it suffices to prove the identity:

$$\int_0^{\infty} dv\, e^{-\lambda v}E_z(g(A_v)) = \frac{1}{\lambda}\int_0^{\infty} dt\, g'(t)E_z[e^{-\lambda\tau_t}] \quad . \tag{6.14}$$

(here, z stands for $X_{\tau_k}(\omega)$ in the previous expressions).

In fact, we now show

$$\int_0^{\infty} dv\, e^{-\lambda v}g(A_v) = \frac{1}{\lambda}\int_0^{\infty} dt\, g'(t)e^{-\lambda\tau_t} \tag{6.15}$$

which, a fortiori, implies (6.14).

Indeed, if we write: $g(a) = \int_0^{a} dt\, g'(t)$, we obtain:

$$\int_0^{\infty} dv\, e^{-\lambda v}g(A_v) = \int_0^{\infty} dv\, e^{-\lambda v}\int_0^{A_v} dt\, g'(t)$$

$$= \int_0^\infty dt \, g'(t) \int_{\tau_t}^\infty dv \, e^{-\lambda v} = \frac{1}{\lambda} \int_0^\infty dt \, g'(t) e^{-\lambda \tau_t}$$

which is precisely the identity (6.15). $\square$

Exercise 6.1 Let $(M_t, t \geq 0)$ be an $\mathbb{R}_+$-valued multiplicative functional of the process X; prove the following generalizations of formulae (6.12) and (6.13):

$$\int_0^\infty dt \, E_x \left[M_t g(A_t - k) \right] = \int_0^\infty dv \, E_x \left[M_{\tau_k} E_{X_{\tau_k}} (M_v g(A_v)) \right]$$

$$= \int_0^\infty dt \, g'(t) E_x \left[M_{\tau_k} E_{X_{\tau_k}} \left(\int_{\tau_t}^\infty dv \, M_v \right) \right]$$

6.3 The case of Lévy processes

We now consider the particular case where (X_t) is a Lévy process, that is a process with homogeneous, independent increments, and we take for (A_t) and g the following:

$$A_t = \int_0^t ds \, \exp(m X_s) \,, \quad \text{and} \quad g(x) = (x^+)^n \,,$$

for some $m \in \mathbb{R}$, and $n > 0$.

We define $Y_k = \exp(X_{\tau_k})$, $y = \exp(x)$, and we denote by $(\tilde{P}_y)_{y \in \mathbb{R}_+}$ the family of laws of the strong Markov process $(Y_k; k \geq 0)$.

We now compute the quantities $G_x(t)$ and $G_x^{(\lambda)}(k)$ in this particular case; we find:

$$G_x(t) = \exp(mnx) e_n(t) = y^{mn} e_n(t) \,,$$

where:

$$e_n(t) = G_0(t) \equiv E_0 \left[\left(\int_0^t ds \, \exp(m X_s) \right)^n \right] \,.$$

On the other hand, we have:

$$\tau_k = \int_0^k \frac{dv}{(Y_v)^m} \quad , \tag{6.16}$$

and formula (6.12) now becomes:

$$G_x^{(\lambda)}(k) = \tilde{E}_y \left[(Y_k)^{mn} \exp(-\lambda \tau_k) \right] e_n^{(\lambda)} \quad , \quad \text{where:} \quad e_n^{(\lambda)} = \int_0^\infty dt e^{-\lambda t} e_n(t) \quad .$$

We may now write both formulae (6.12) and (6.13) as follows.

Proposition 6.2 *With the above notation, we have:*

$$G_x^{(\lambda)}(k) \underset{(i)}{=} \tilde{E}_y \left[(Y_k)^{mn} \exp(-\lambda \tau_k) \right] e_n^{(\lambda)} \underset{(ii)}{=} \frac{n}{\lambda} \int_k^\infty dv (v-k)^{n-1} \tilde{E}_y [e^{-\lambda \tau_v}] \tag{6.17}$$

In the particular case $n = 1$, this double equality takes a simpler form: indeed, in this case, we have

$$e_1^{(\lambda)} = \int_0^\infty dt \, e^{-\lambda t} e_1(t) = \int_0^\infty dt \, e^{-\lambda t} \int_0^t ds \exp(s \, \varphi(m)) \quad ,$$

where φ is the Lévy exponent of X. It is now elementary to obtain, for $\lambda > \varphi(m)$, the formula: $e_1^{(\lambda)} = \dfrac{1}{\lambda(\lambda - \varphi(m))}$, and, therefore, for $n = 1$, the formulae (6.17) become

$$\lambda G_x^{(\lambda)}(k) = \tilde{E}_y \left[(Y_k)^m \exp(-\lambda \tau_k) \right] \frac{1}{(\lambda - \varphi(m))} = \int_k^\infty dv \, \tilde{E}_y [\exp(-\lambda \tau_v)] \quad .$$

$$\tag{6.18}$$

6.4 Application to Brownian motion

We now assume that: $X_t = B_t + \nu t$, $t \geq 0$, with (B_t) a Brownian motion, and $\nu \geq 0$, and we take $m = 2$, which implies:

$$A_t = \int_0^t ds \exp(2X_s) \ .$$

In this particular situation, the process $(Y_k, k \geq 0)$ is now the Bessel process with index ν, or dimension $\delta_\nu = 2(1 + \nu)$. We denote by $P_y^{(\nu)}$ the law of this process, when starting at y, and we write simply $P^{(\nu)}$ for $P_1^{(\nu)}$. Hence, for example, $P^{(0)}$ denotes the law of the 2-dimensional Bessel process, starting from 1. We now recall the Girsanov relation, which was already used in Chapter 5, formula (5.8):

$$P_y^{(\nu)}\big|_{\mathcal{R}_t} = \left(\frac{R_t}{y}\right)^\nu \exp\left(-\frac{\nu^2}{2}\tau_t\right) \cdot P_y^{(0)}\big|_{\mathcal{R}_t}, \quad \text{where } \tau_t = \int_0^t \frac{ds}{R_s^2} \ . \quad (6.19)$$

In Chapter 5, we used the notation H_t for τ_t; $(R_t, t \geq 0)$ denotes, as usual, the coordinate process on Ω_+^*, and $\mathcal{R}_t = \sigma\{R_s, s \leq t\}$. The following Lemma is now an immediate consequence of formula (6.19).

Lemma 6.1 *For every $\alpha \in \mathbb{R}$, for every $\nu \geq 0$, and $\lambda \geq 0$, we have, if we denote: $\mu = \sqrt{2\lambda + \nu^2}$,*

$$E^{(\nu)}\left[(R_k)^\alpha \exp(-\lambda\tau_k)\right] = E^{(0)}\left[(R_k)^{\alpha+\nu} \exp\left(-\frac{\mu^2}{2}\tau_k\right)\right] = E^{(\mu)}\left[R_k^{\alpha+\nu-\mu}\right] \quad (6.20)$$

We are now able to write the formulae (6.17) in terms of the moments of Bessel processes.

Proposition 6.3 *We now write simply $G^{(\lambda)}(k)$ for $G_0^{(\lambda)}(k)$, and we introduce the notation:*

$$H_\mu(\alpha; s) = E^{(\mu)}((R_s)^\alpha) \ . \quad (6.21)$$

Then, we have:

$$G^{(\lambda)}(k) \underset{(i)}{=} H_\mu(2n + \nu - \mu; k)e_n^{(\lambda)} \underset{(ii)}{=} \frac{n}{\lambda}\int_k^\infty dv(v - k)^{n-1} H_\mu(\nu - \mu; v) \quad (6.22)$$

which, in the particular case $n = 1$, simplifies, with the notation: $\delta_\nu = 2(1 + \nu)$, to:

$$\lambda G^{(\lambda)}(k) \underset{(i)}{=} H_\mu(2 + \nu - \mu; k)\frac{1}{(\lambda - \delta_\nu)} \underset{(ii)}{=} \int_k^\infty dv H_\mu(\nu - \mu; v) \ . \quad (6.23)$$

It is now clear, from formula (6.22) that in order to obtain a closed form formula for $G^{(\lambda)}(k)$, it suffices to be able to compute explicitly $H_\mu(\alpha, k)$ and $e_n^{(\lambda)}$. In fact, once $H_\mu(\alpha; k)$ is computed for all admissible values of α and k, by taking $k = 0$ in formula (6.22) (ii), we obtain:

$$e_n^{(\lambda)} = \frac{n}{\lambda} \int_0^\infty dv \ v^{n-1} H_\mu(\nu - \mu; v) \ , \tag{6.24}$$

from which we shall deduce formula (6.3) for $\lambda e_n^{(\lambda)} \equiv E\left[(A_{T_\lambda}^{(\nu)})^n\right]$.

We now present the quickest way, to our knowledge, to compute $H_\mu(\alpha; k)$. In order to compute this quantity, we find it interesting to introduce the laws Q_z^δ of the square Bessel process $(\Sigma_u, u \geq 0)$ of dimension δ, starting from z, for $\delta > 0$, and $z > 0$, because of the additivity property of this family (see Chapter 2, Theorem 2.3).

We then have the following

Proposition 6.4 *For $z > 0$, and for every γ such that: $0 < \gamma < \mu + 1$, we have:*

$$\frac{1}{z^\gamma} H_\mu\left(-2\gamma; \frac{1}{2z}\right) \underset{(i)}{=} Q_z^{\delta_\mu}\left(\frac{1}{(\Sigma_{1/2})^\gamma}\right) \underset{(ii)}{=} \frac{1}{\Gamma(\gamma)} \int_0^1 du \ e^{-zu} u^{\gamma-1}(1-u)^{\mu-\gamma}$$
$$\tag{6.25}$$

PROOF:

a) Formula (6.25)(i) is a consequence of the invariance property of the laws of Bessel processes by time-inversion;

b) We now show how to deduce formula (6.25)(ii) from (6.25)(i). Using the elementary identity:

$$\frac{1}{r^\gamma} = \frac{1}{\Gamma(\gamma)} \int_0^\infty dt \ e^{-rt} t^{\gamma-1} \ ,$$

we obtain:

$$Q_z^{\delta_\mu}\left(\frac{1}{(\Sigma_{1/2})^\gamma}\right) = \frac{1}{\Gamma(\gamma)} \int_0^\infty dt \ t^{\gamma-1} Q_z^{\delta_\mu}(e^{-t\Sigma_{1/2}})$$

and the result now follows from the general formula:

$$Q_z^\delta \left(\exp(-\alpha \Sigma_s) \right) = \frac{1}{(1 + 2\alpha s)^{\delta/2}} \exp \left(-z \frac{\alpha}{1 + 2\alpha s} \right) , \tag{6.26}$$

which we use with $\alpha = t$, and $s = 1/2$.   □

Remark: Formula (6.26) is easily deduced from the additivity property of the family (Q_z^δ) (see Revuz-Yor [81], p. 411).   □

We now show how formulae (6.2) and (6.3) are consequences of formula (6.25):

- firstly, we apply formula (6.22)(i), together with formula (6.25)(ii), with $\gamma = \frac{\mu - \nu}{2} - n$, and $z = x$. Formula (6.2) then follows after making the change of variables: $u = \frac{t}{x}$ in the integral in formula (6.25);
- secondly, we take formula (6.22)(ii) with $k = 0$, which implies:

$$E \left[(A_{T_\lambda}^{(\nu)})^n \right] = n \int_0^\infty dv \; v^{n-1} H_\mu(\nu - \mu; v) ,$$

and we then obtain formula (6.3) by replacing in the above integral $H_\mu(\nu - \mu; v)$ by its value given by (6.25)(ii), with $\gamma = \frac{\mu - \nu}{2}$.

In fact, when we analyze the previous arguments in detail, we obtain a representation of the r.v. $A_{T_\lambda}^{(\nu)}$ as the ratio of a beta variable to a gamma variable, both variables being independent; such analysis also provides us with some very partial explanation of this independence property. Precisely, we have obtained the following result.

Theorem 6.3 *1. The law of the r.v. $A_{T_\lambda}^{(\nu)}$ satisfies*

$$A_{T_\lambda}^{(\nu)} \overset{\text{(law)}}{=} \frac{Z_{1,a}}{2Z_b} , \qquad \text{where } a = \frac{\mu + \nu}{2} \quad \text{and } b = \frac{\mu - \nu}{2} \tag{6.27}$$

and where $Z_{\alpha,\beta}$, resp. Z_b, denotes a beta variable with parameters α and β, resp. a gamma variable with parameter b, and both variables on the right hand side of (6.27) are independent.

2. More generally, we obtain:

$$\left(A_{T_\lambda}^{(\nu)}; \frac{Z_a}{Z_b} \exp(2B_{T_\lambda}^{(\nu)}) \right) \overset{\text{(law)}}{=} \left(\frac{Z_1}{2(Z_1 + Z_a)Z_b}; \frac{Z_a}{Z_b} \exp(2B_{T_\lambda}^{(\nu)}) \right) \tag{6.28}$$

where Z_1, Z_a, Z_b are three independent gamma variables, with respective parameters $1, a, b$, and these variables are also assumed to be independent of B and T_λ.

Remark: Our aim in establishing formula (6.28) was to try and understand better the factorization which occurs in formula (6.27), but, at least at first glance, formula (6.28) does not seem to be very helpful.

PROOF OF THE THEOREM:

a) From formula (6.24), if we take n sufficiently small, we obtain:

$$
\begin{aligned}
E\left[(A_{T_\lambda}^{(\nu)})^n\right] &= \int_0^\infty dv\; nv^{n-1} H_\mu(\nu - \mu; v) \\
&= \int_0^\infty \frac{dy}{y} n \left(\frac{1}{2y}\right)^n y^b \left\{\frac{1}{y^b} H_\mu\left(-2b; \frac{1}{2y}\right)\right\}, \quad \text{where } b = \frac{\mu - \nu}{2} \\
&= \int_0^\infty \frac{dy}{y} n \left(\frac{1}{2y}\right)^n y^b Q_y^{\delta_\mu}\left(\frac{1}{\Sigma_{1/2}^b}\right), \quad \text{from } (6.25)(i) \\
&= \int_0^\infty \frac{dy}{y} n \left(\frac{1}{2y}\right)^n E\left[\exp -y Z_{(b,a+1)}\right] c_{\mu,\nu}, \quad \text{from } (6.25)(ii).
\end{aligned}
$$

In the sequel, the constant $c_{\mu,\nu}$ may vary, but shall never depend on n. For simplicity, we now write Z instead of $Z_{(b,a+1)}$, and we obtain, after making the change of variables: $y = z/Z$:

$$
\begin{aligned}
E\left[(A_{T_\lambda}^{(\nu)})^n\right] &= c_{\mu,\nu} E\left[\int_0^\infty \frac{dz}{z} n \left(\frac{Z}{2z}\right)^n \left(\frac{z}{Z}\right)^b \exp(-z)\right] \\
&= c_{\mu,\nu} E\left[n Z^{n-1} \frac{1}{Z^{b-1}}\right] \int_0^\infty dz \left(\frac{1}{2z}\right)^n z^{b-1} e^{-z},
\end{aligned}
$$

and, after performing an integration by parts in the first expectation, we obtain:

$$
E\left[(A_{T_\lambda}^{(\nu)})^n\right] = E\left[(Z_{1,a})^n\right] E\left[\left(\frac{1}{2Z_b}\right)^n\right] \tag{6.29}
$$

which implies (6.27).

b) We take up the same method as above, that is: we consider

$$E\left[(A_{T_\lambda}^{(\nu)})^\alpha \exp(\beta B_{T_\lambda}^{(\nu)})\right] \quad , \quad \text{for small } \alpha \text{ and } \beta\text{'s.}$$

Applying Cameron-Martin's absolute continuity relationship between Brownian motion and Brownian motion with drift, we find:

$$
\begin{aligned}
E\left[(A_{T_\lambda}^{(\nu)})^\alpha \exp(\beta B_{T_\lambda}^{(\nu)})\right] &\\
= \lambda \int_0^\infty dt \exp&\left\{-\left(\lambda + \frac{\nu^2}{2}\right)t\right\} E\left[(A_t^{(0)})^\alpha \exp(\beta + \nu)B_t\right]\\
= \lambda \int_0^\infty dt\, e^{-\theta t} & E\left[(A_t^{(\beta+\nu)})^\alpha\right] = \frac{\lambda}{\theta} E\left[(A_{T_\theta}^{(\beta+\nu)})^\alpha\right] \quad ,
\end{aligned}
$$

where $\theta = \lambda + \frac{\nu^2}{2} - \frac{(\beta+\nu)^2}{2} = \lambda - \frac{\beta^2}{2} - \beta\nu$.

We now remark that $\mu' \overset{\text{def}}{=} \sqrt{2\theta + (\beta+\nu)^2}$ is in fact equal to $\mu = \sqrt{2\lambda + \nu^2}$, so that we may write, with the help of formula (6.29):

$$E\left[(A_{T_\lambda}^{(\nu)})^\alpha \exp(\beta B_{T_\lambda}^{(\nu)})\right] = \left(\frac{\lambda}{\theta}\right) E\left[\left(Z_{1,a+\frac{\beta}{2}}\right)^\alpha\right] E\left[\left(\frac{1}{2Z_{b-\frac{\beta}{2}}}\right)^\alpha\right] \quad . \tag{6.30}$$

Now, there exist constants C_1 and C_2 such that:

$$
\begin{aligned}
E\left[(Z_{1,a+\frac{\beta}{2}})^\alpha\right] &= E\left[(Z_{1,a})^\alpha (1 - Z_{1,a})^{\beta/2}\right] C_1\\
E\left[\left(2Z_{b-\frac{\beta}{2}}\right)^{-\alpha}\right] &= E\left[(2Z_b)^{-\alpha}(Z_b)^{-\beta/2}\right] C_2
\end{aligned}
$$

and it is easily found that: $C_1 = \frac{a+\beta/2}{a}$ and $C_2 = \frac{\Gamma(b)}{\Gamma(b-\frac{\beta}{2})}$. Furthermore, we now remark that, by taking simply $\alpha = 0$ in formula (6.30):

$$E\left[\exp\left(\beta B_{T_\lambda}^{(\nu)}\right)\right] = \frac{\lambda}{\theta} \quad .$$

Hence, we may write formula (6.30) as:

$$E\left[(A_{T_\lambda}^{(\nu)})^\alpha \exp(\beta B_{T_\lambda}^{(\nu)})\right] \frac{1}{C_1 C_2} = E\left[\left(\frac{Z_{1,a}}{2Z_b}\right)^\alpha \left(\frac{1-Z_{1,a}}{Z_b}\right)^{\beta/2} \exp(\beta B_{T_\lambda}^{(\nu)})\right].$$

Now, since: $\dfrac{1}{C_1 C_2} = E\left[\left(\dfrac{Z_{a,1}}{Z_b}\right)^{\beta/2}\right]$, we deduce from the above identity that:

$$\left(A_{T_\lambda}^{(\nu)}; \left(\frac{Z_{a,1}}{Z_b}\right)^{1/2} \exp(B_{T_\lambda}^{(\nu)})\right) \overset{(\text{law})}{=} \left(\frac{Z_{1,a}}{2Z_b}; \left(\frac{1-Z_{1,a}}{Z_b}\right)^{1/2} \exp(B_{T_\lambda}^{(\nu)})\right)$$

$$\overset{(\text{law})}{=} \left(\frac{1-Z_{a,1}}{2Z_b}; \left(\frac{Z_{a,1}}{Z_b}\right)^{1/2} \exp\left(B_{T_\lambda}^{(\nu)}\right)\right)$$

from which we easily obtain (6.28) thanks to the beta-gamma relationships.
□

As a verification, we now show that formula (6.2) may be recovered simply from formula (6.27); it is convenient to write formula (6.2) in the equivalent form:

$$\frac{1}{x^{b-n}} E\left[\left(\left\{\frac{Z_{1,a}}{Z_b} - \frac{1}{x}\right\}^+\right)^n\right] = \frac{E\left[\left(\frac{Z_{1,a}}{Z_b}\right)^n\right]}{\Gamma(b-n)} \int_0^1 dt \, e^{-xt} t^{b-n-1} (1-t)^{n+a}$$

(6.31)

for $x > 0$, $n < b$, and $a > 0$.

We now obtain the more general formula

Proposition 6.5 *Let $Z_{\alpha,\beta}$ and Z_γ be two independent random variables, which are, respectively a beta variable with parameters (α, β) and a gamma variable with parameter γ. Then, we have, for every $x > 0$, and $n < \gamma$:*

$$E\left[\left\{\left(\frac{Z_{\alpha,\beta}}{Z_\gamma} - \frac{1}{x}\right)^+\right\}^n\right] = \frac{x^{\gamma-n}}{\Gamma(\gamma) B(\alpha, \beta)} \int_0^1 du \, e^{-xu} u^{\gamma-n-1} (1-u)^{\beta+n} \ldots$$

$$\ldots \int_0^1 dw \, (u + w(1-u))^{\alpha-1} w^n (1-w)^{\beta-1} . (6.32)$$

In the particular case $\alpha = 1$, formula (6.32) simplifies to:

$$E\left[\left\{\left(\frac{Z_{1,\beta}}{Z_\gamma} - \frac{1}{x}\right)^+\right\}^n\right] = \frac{x^{\gamma-n}}{\Gamma(\gamma)} \left(\int_0^1 du \, e^{-xu} u^{\gamma-n-1} (1-u)^{\beta+n}\right) (\beta B(n+1, \beta))$$

(6.33)

which is precisely formula (6.31), taken with $a = \beta$, and $b = \gamma$.

PROOF: We remark that we need only integrate upon the subset of the probability space $\{1 \geq Z_{\alpha,\beta} \geq \frac{1}{x} Z_\gamma\}$, and after conditioning on Z_γ, we integrate with respect to the law of $Z_{\alpha,\beta}$ on the random interval $[\frac{1}{x} Z_\gamma; 1]$. This gives:

$$E\left[\left\{\left(\frac{Z_{\alpha,\beta}}{Z_\gamma} - \frac{1}{x}\right)^+\right\}^n\right] = E\left[\frac{1}{Z_\gamma^n}\left\{\left(Z_{\alpha,\beta} - \frac{Z_\gamma}{x}\right)^+\right\}^n\right]$$

$$= E\left[\frac{1}{Z_\gamma^n}\frac{1}{B(\alpha,\beta)}\int_{\frac{Z_\gamma}{x}}^{1} du\, u^{\alpha-1}\left(u - \frac{Z_\gamma}{x}\right)^n(1-u)^{\beta-1}\right]$$

and the rest of the computation is routine. $\square$

We now consider some particularly interesting subcases of formula (6.27)

Theorem 6.4 *Let U be a uniform variable on $[0,1]$, and $\sigma \overset{\text{def}}{=} \inf\{t : B_t = 1\}$.*

1) For any $\nu \in [0,1[$, we have:

$$A_{T_{2(1-\nu)}}^{(\nu)} \overset{(\text{law})}{=} \frac{U}{2Z_{1-\nu}} \tag{6.34}$$

In particular, we have, taking $\nu = 0$, and $\nu = \frac{1}{2}$, respectively:

$$\int_0^{T_1} ds\,\exp(\sqrt{2}B_s) \overset{(\text{law})}{=} \frac{U}{2Z_1} \quad\text{and}\quad \int_0^{T_1} ds\,\exp(2B_s + s) \overset{(\text{law})}{=} U\sigma \tag{6.35}$$

where, as usual, the variables which appear on the right-hand sides of (6.34) and (6.35) are assumed to be independent.

2) For any $\nu \geq 0$,

$$A_{T_{\nu+\frac{1}{2}}}^{(\nu)} \overset{(\text{law})}{=} Z_{1,\nu+\frac{1}{2}}\,\sigma\,. \tag{6.36}$$

PROOF: The different statements follow immediately from formula (6.27), once one has remarked that:

$$Z_{1,1} \overset{(\text{law})}{=} U \quad\text{and}\quad \frac{1}{2Z_{1/2}} \overset{(\text{law})}{=} \frac{1}{N^2} \overset{(\text{law})}{=} \sigma\,,$$

where N is a centered Gaussian variable with variance 1. $\square$

6.5 A discussion of some identities

(6.5.1)   Formula (6.25)(ii) might also have been obtained by using the explicit expression of the semi-group of the square of a Bessel process (see, for example, [81] p. 411, Corollary (1.4)). With this approach, one obtains the following formula:

$$\frac{1}{z^\gamma} H_\mu\left(-2\gamma; \frac{1}{2z}\right) \equiv Q_z^{\delta_\mu}\left(\frac{1}{(\Sigma_{1/2})^\gamma}\right) = \exp(-z)\frac{\Gamma(\alpha)}{\Gamma(\beta)}\Phi(\alpha, \beta; z) \qquad (6.37)$$

where $\alpha = -\gamma + 1 + \mu$, $\beta = 1 + \mu$, and $\Phi(\alpha, \beta; z)$ denotes the confluent hypergeometric function with parameters α and β.

With the help of the following classical relations (see Lebedev [63], p. 266–267):

$$
\begin{cases}
\text{(i)} \qquad\qquad \Phi(\alpha, \beta; z) = e^z \Phi(\beta - \alpha, \beta; -z) \\[2mm]
\qquad\qquad\qquad\qquad\text{and} \\[2mm]
\text{(ii) } \Phi(\beta - \alpha, \beta; z) = \dfrac{\Gamma(\beta)}{\Gamma(\alpha)\Gamma(\beta - \alpha)} \displaystyle\int_0^1 dt\, e^{-zt} t^{(\beta - \alpha)-1}(1 - t)^{\alpha - 1},
\end{cases}
\qquad (6.38)
$$

one may obtain formula (6.25)(ii) as a consequence of (6.37).

(6.5.2)   The recurrence formula (6.22)(ii) may be written, after some elementary transformations, in the form:

$$x^\alpha H_\mu\left(2\alpha; \frac{1}{2x}\right) e_n^{(\lambda)} = \frac{n}{\lambda 2^{n-1}} \int_0^1 dw\, w^{-\alpha-1}(1 - w)^{n-1}(xw)^\beta H_\mu\left(2\beta; \frac{1}{2wx}\right)$$

$$(6.39)$$

where, now, we take $\alpha = n + \frac{\nu - \mu}{2}$, and $\beta = \frac{\nu - \mu}{2} \equiv \alpha - n$.

Assuming that formula (6.25)(ii) is known, the equality (6.39) is nothing but an analytic translation of the well-known algebraic relation between beta variables

$$Z_{a,b+c} \overset{(\text{law})}{=} Z_{a,b} Z_{a+b,c} \qquad (6.40)$$

for the values of the parameters: $a = -\alpha$, $b = n$, $c = 1 + \mu + \beta$, and $Z_{p,q}$ denotes a beta variable with parameters (p, q), and the two variables on the right-hand side of (6.40) are assumed to be independent. In terms of confluent hypergeometric functions, the equality (6.39) translates into the identity:

$$\Phi(\alpha, \gamma; z) = \frac{\Gamma(\gamma)}{\Gamma(\beta)\Gamma(\gamma - \beta)} \int_0^1 dt \, t^{\beta-1}(1 - t)^{\gamma-\beta-1}\Phi(\alpha, \beta; zt) \qquad (6.41)$$

for $\gamma > \beta$ (see Lebedev [63], p. 278).

(6.5.3)   The relations (6.39) and (6.41) may also be understood in terms of the semigroups $(Q_t^{\delta}; t \geq 0)$ and $(Q_t^{\delta'}; t \geq 0)$ of squares of Bessel processes with respective dimensions δ and δ', via the intertwining relationship:

$$Q_t^{\delta} M_{k',k} = M_{k',k} Q_t^{\delta'} \ , \qquad (6.42)$$

where: $0 < \delta' < \delta$, $k' = \frac{\delta'}{2}$, $k = \frac{\delta - \delta'}{2}$, and $M_{a,b}$ is the "multiplication" Markov kernel which is defined by:

$$M_{a,b}f(x) = E[f(xZ_{a,b})] \ , \quad \text{where } f \in b(\mathcal{B}(\mathbb{R}_+)) \qquad (6.43)$$

(for a discussion of such intertwining relations, which are closely linked with beta and gamma variables, see Yor [99]).

(6.5.4)   Finally, we close up this discussion with a remark which relates the recurrence formula (6.22)(ii), in which we assume n to be an integer, to the uniform integrability property of the martingales

$$M_k^{(p)} \stackrel{\text{def}}{=} R_k^{2+p} - 1 - c_p^{(\mu)} \int_0^k ds \, R_s^p \ , \quad k \geq 0, \text{ under } P^{(\mu)}, \qquad (6.44)$$

for: $-2\mu < 2 + p < 0$, with $c_p^{(\mu)} = (2 + p)\left(\mu + \frac{2+p}{2}\right)$.

Once this uniform integrability property, under the above restrictions for p, has been obtained, one gets, using the fact that: $R_\infty^{2+p} = 0$, the following relation:

$$E^{(\mu)}(R_k^{2+p}) = -c_p^{(\mu)} E^{(\mu)}\left(\int_k^\infty ds \, R_s^p\right) \ ,$$

and, using this relation recurrently, one obtains formula (6.22)(ii) with the following expression for the constant $\lambda e_n^{(\lambda)} \equiv E\left((A_{T_\lambda}^{(\nu)})^n\right)$

$$E\left[(A_{T_\lambda}^{(\nu)})^n\right] = (-1)^n \frac{n!}{\prod\limits_{j=0}^{n-1} c_{2j+\nu-\mu}^{(\mu)}} \tag{6.45}$$

and an immediate computation shows that formulae (6.45) and (6.4) are identical.

Comments on Chapter 6

Whereas, in Chapter 5, some studies of a continuous determination of the logarithm along the planar Brownian trajectory have been made, we are interested here in the study of the laws of exponential functionals of Brownian motion, or Brownian motion with drift.

The origin of the present study comes from Mathematical finance; the so-called financial Asian options take into account the past history of the market, hence the introduction of the arithmetic mean of the geometric Brownian motion. A thorough discussion of the motivation from Mathematical finance is made in Geman-Yor [47]. The results in paragraph 6.1 are taken from Yor [102].

The developments made in paragraphs 6.2 and 6.3 show that there are potential extensions to exponential functionals of a large class of Lévy processes. However, the limitation of the method lies in the fact that, if (X_t) is a Lévy process, and $(R(u), u \geq 0)$ is defined by:

$$\exp(X_t) = R\left(\int_0^t ds \exp(X_s)\right) \ ,$$

then, the semi-group of R is only known explicitly in some particular cases, but a class of examples has been studied, in joint work with Ph. Carmona and F. Petit [24].

In paragraph 6.4, a simple description of the law of the variable $A_{T_\lambda}^{(\nu)}$ is obtained; it would be nice to be able to explain the origin of the beta variable, resp.: gamma variable, in formula (6.27), from, possibly, a path decomposition. In paragraph 6.5, a discussion of the previously obtained formulae in terms of confluent hypergeometric functions is presented.

Had we chosen, for our computations, the differential equations approach which is closely related to the Feynman-Kac formula, these functions would have immediately appeared. However, throughout this chapter, and in related publications (Geman-Yor [46], [47], and Yor [102]), we have preferred to use some adequate change of probability, and Girsanov theorem.

The methodology used in this Chapter helped to unify certain computations for Asian, Parisian and barrier options (see [105]).

The Springer-Finance volume [104] gathers ten papers dealing, in a broad sense, with Asian options.

Chapter 7

Some asymptotic laws for multidimensional BM

In this chapter, we first build upon the knowledge gained in Chapter 5 about the asymptotic windings of planar BM around one point, together with the Kallianpur-Robbins ergodic theorem for planar BM, to extend Spitzer's theorem:

$$\frac{2\theta_t}{\log t} \xrightarrow[t\to\infty]{\text{(law)}} C_1$$

into a multidimensional result for the winding numbers $(\theta_t^1, \theta_1^2, \ldots, \theta_t^n)$ of planar BM around n points (all notations which may be alluded to, but not defined in this chapter are found in Pitman-Yor [75]).

This study in the plane may be extended one step further by considering BM in $\mathbb{R}^3$ and seeking asymptotic laws for its winding numbers around a finite number of oriented straight lines, or, even, certain unbounded curves (Le Gall-Yor [62]).

There is, again, a more general set-up for which such asymptotic laws may be obtained, and which allows to unify the previous studies: we consider a finite number $(B^1, B^2, \ldots, B^m)$ of jointly Gaussian, "linearly correlated", planar Brownian motions, and the winding numbers of each of them around z^j, where $\{z^j; 1 \leq j \leq n\}$ are a finite set of points.

In the last paragraph, some asymptotic results for Gauss linking numbers relative to one BM, or two independent BM's, with values in $\mathbb{R}^3$ are presented.

7.1 Asymptotic windings of planar BM around n points

In Chapter 5, we presented Spitzer's result

$$\frac{2\theta_t}{\log t} \xrightarrow[t \to \infty]{\text{(law)}} C_1 \ . \tag{7.1}$$

This may be extended as follows:

$$\frac{2}{\log t} \left(\theta_t^{r,-}, \theta_t^{r,+}, \int_0^t ds \, f(Z_s) \right) \xrightarrow[t \to \infty]{\text{(law)}} \left(\int_0^\sigma d\gamma_s 1_{(\beta_s \leq 0)}, \int_0^\sigma d\gamma_s 1_{(\beta_s \geq 0)}, \left(\frac{\bar{f}}{2\pi} \right) \ell_\sigma \right) \tag{7.2}$$

where: $\theta_t^{r,-} = \int_0^t d\theta_s 1_{(|Z_s| \leq r)}$, $\theta_t^{r,+} = \int_0^t d\theta_s 1_{(|Z_s| \geq r)}$, $f : \mathbb{C} \to \mathbb{R}$ is inte-

grable with respect to Lebesgue measure, $\bar{f} = \iint_{\mathbb{C}} dx \, dy \, f(z)$, β and γ are

two independent real Brownian motions, starting from 0, $\sigma = \inf\{t : \beta_t = 1\}$, and ℓ_σ is the local time at level 0, up to time σ of β (for a proof of (7.2), see Messulam-Yor [65] and Pitman-Yor [75]).

The result (7.2) shows in particular that Spitzer's law (7.1) takes place jointly with the Kallianpur-Robbins law (which is the convergence in law of the third component on the left-hand side of (7.2) towards an exponential variable; see, e.g., subparagraph (4.3.2), case 3)).

A remarkable feature in (7.2) is that the right-hand side does not depend on r. The following Proposition provides an explanation, which will be a key for the extension of (7.1) to the asymptotic study of the winding numbers with respect to a finite number of points.

Proposition 7.1 *Let* $\varphi(z) = (f(z); g(z))$ *be a function from* $\mathbb{C}$ *to* $\mathbb{R}^2$ *such that:*

$$\iint dx \, dy \, |\varphi(z)|^2 \equiv \iint dx \, dy \, \{(f(z))^2 + (g(z))^2\} < \infty$$

Then, the following quantity:

$$\frac{1}{\sqrt{\log t}} \int_0^t \varphi(Z_s) \cdot dZ_s \equiv \frac{1}{\sqrt{\log t}} \int_0^t (dX_s f(Z_s) + dY_s g(Z_s))$$

converges in law, as $t \to \infty$, *towards:*

$$\sqrt{k_\varphi}\ \Gamma_{\frac{1}{2}\ell_\sigma}$$

where $k_\varphi \equiv \frac{1}{2\pi}\iint dx\,dy|\varphi(z)|^2$, *and* $(\Gamma_t, t \geq 0)$ *is a 1-dimensional BM. This convergence in law takes place jointly with (7.2), and* Γ, β, γ *are independent.*

For this Proposition, see Messulam-Yor [65] and Kasahara-Kotani [55]. Proposition 7.1 gives an explanation for the absence of the radius r on the right-hand side of (7.2); more precisely, the winding number in the annulus:

$$\{z : r \leq |z| \leq R\}\ ,\quad \text{for } 0 < r < R < \infty\ ,$$

is, roughly, of the order of $\sqrt{\log t}$, and, therefore:

$$\frac{1}{\log t}\theta_t^{r,R} \equiv \frac{1}{\log t}\int_0^t d\theta_s 1_{(r \leq |Z_s| \leq R)} \xrightarrow[t \to \infty]{(P)} 0\ .$$

We now consider $\theta_t^1, \theta_t^2, \ldots, \theta_t^n$, the winding numbers of $(Z_u, u \leq t)$ around each of the points $z^1, z^2, \ldots, z^n$. Just as before, we separate θ_t^j into $\theta_t^{j,-}$ and $\theta_t^{j,+}$, where, for some $r_j > 0$, we define:

$$\theta_t^{j,-} = \int_0^t d\theta_s^j 1_{(|Z_s - z^j| \leq r_j)}\quad \text{and}\quad \theta_t^{j,+} = \int_0^t d\theta_s^j 1_{(|Z_s - z^j| \geq r_j)}$$

Another application of Proposition 7.1 entails:

$$\frac{1}{\log t}\left|\theta_t^{i,+} - \theta_t^{j,+}\right| \xrightarrow[t \to \infty]{(P)} 0\ , \tag{7.3}$$

so that it is now quite plausible, and indeed it is true, that:

$$\frac{2}{\log t}(\theta_t^1, \ldots, \theta_t^n) \xrightarrow[t \to \infty]{\text{(law)}} \left(W_1^- + W^+, W_2^- + W^+, \ldots, W_n^- + W^+\right) \tag{7.4}$$

Moreover, the asymptotic random vector: $\left(W_1^-, W_2^-, \ldots, W_n^-, W^+\right)$ may be represented as

$$(L_T(U)C_k\ (1 \leq k \leq n)\ ;\ V_T) \tag{7.5}$$

where $(U_t, t \geq 0)$ is a reflecting BM, $T = \inf\{t : U_t = 1\}$, $L_T(U)$ is the local time of U at 0, up to time T, $(V_t, t \geq 0)$ is a one-dimensional BM, starting from 0, which is independent of U, and $(C_k; 1 \leq k \leq n)$ are independent Cauchy variables with parameter 1, which are also independent of U and V. The representation (7.5) agrees with (7.2) as one may show that:

$$\left(\int_0^\sigma d\gamma_s 1_{(\beta_s \leq 0)}; \int_0^\sigma d\gamma_s 1_{(\beta_s \geq 0)}; \frac{1}{2} \ell_\sigma \right) \overset{\text{(law)}}{=} (L_T(U)C_1; V_T; L_T(U))$$

essentially by using, as we already did in paragraph 4.1, the well-known representation:

$$\beta_t^+ = U_{\int_0^t ds 1_{(\beta_s \geq 0)}} \qquad , \qquad t \geq 0 \ .$$

From the formula for the characteristic function:

$$E\left[\exp i\left(\alpha V_T + \beta L_T(U)C_1\right)\right] = \left(\operatorname{ch}\alpha + |\beta|\frac{\operatorname{sh}\alpha}{\alpha}\right)^{-1} ,$$

(which may be derived directly, or considered as a particular case of the first formula in subparagraph (3.3.2)), it is easy to obtain the multidimensional explicit formula:

$$E\left[\exp\left(i\sum_{k=1}^n \alpha_k W_k\right)\right] = \left(\operatorname{ch}\left(\sum_{k=1}^n \alpha_k\right) + \frac{\sum|\alpha_k|}{\sum\alpha_k}\operatorname{sh}\left(\sum_{k=1}^n \alpha_k\right)\right)^{-1} , \quad (7.6)$$

where we have denoted W_k for $W_k^- + W^+$.

Formula (7.6) shows clearly, if needed, that each of the W_k's is a Cauchy variable with parameter 1, and that these Cauchy variables are stochastically dependent, in an interesting manner, which is precisely described by the representation (7.5).

The following asymptotic residue theorem may now be understood as a global summary of the preceding results.

Theorem 7.1

1) *Let f be holomorphic in $\mathbb{C} \setminus \{z^1, \ldots, z^n\}$, and let Γ be an open, relatively compact set such that: $\{z^1, \ldots, z^n\} \subset \Gamma$.*

   Then, one has:

$$\frac{2}{\log t} \int_0^t f(Z_s) 1_\Gamma(Z_s) dZ_s \xrightarrow[t\to\infty]{\text{(law)}} \sum_{j=1}^n Res(f, z^j)(L_T(U) + iW_j^-)$$

2) If, moreover, f is holomorphic at infinity, and $\lim\limits_{z \to \infty} f(z) = 0$, then:

$$\frac{2}{\log t} \int_0^t f(Z_s) dZ_s$$

$$\xrightarrow[t \to \infty]{(law)} \sum_{j=1}^n Res(f, z^j)(L_T(U) + iW_j^-) + Res(f, \infty)(L_T(U) - 1 + iW^+)$$

7.2 Windings of BM in $\mathbb{R}^3$

We define the winding number θ_t^D of $(B_u, u \le t)$, a 3-dimensional BM around an oriented straight line D as the winding number of the projection of B on a plane orthogonal to D. Consequently, if $D_1, \ldots, D_n$ are parallel, the preceding results apply. If D and D' are not parallel, then:

$$\frac{1}{\log t} \theta_t^D \quad \text{and} \quad \frac{1}{\log t} \theta_t^{D'}$$

are asymptotically independent, since both winding numbers are obtained, to the order of $\log t$, by only considering the amount of winding made by $(B_u, u \le t)$ as it wanders within cones of revolution with axes D, resp: D', the aperture of which we can choose as small as we wish. Therefore, these cones may be taken to be disjoint (except possibly for a common vertex). This assertion is an easy consequence of the more precise following statement: consider $B \equiv (X, Y, Z)$ a Brownian motion in $\mathbb{R}^3$, such that $B_0 \notin D_* \equiv \{x = y = 0\}$. To a given Borel function $f : \mathbb{R}_+ \to \mathbb{R}_+$, we associate the volume of revolution:

$$\Gamma^f \equiv \left\{ (x, y, z) : (x^2 + y^2)^{1/2} \le f(|z|) \right\} ,$$

and we define:

$$\theta_t^f = \int_0^t d\theta_s 1_{(B_s \in \Gamma^f)} .$$

We have the following

Theorem 7.2 *If* $\dfrac{\log f(\lambda)}{\log \lambda} \xrightarrow[t \to \infty]{} a$, *then:* $\dfrac{2\theta_t^f}{\log t} \xrightarrow[t \to \infty]{\text{(law)}} \displaystyle\int_0^\sigma d\gamma_u 1_{(\beta_u \leq aS_u)}$ *where* β *and* γ *are two independent real-valued Brownian motions,* $S_u = \sup\limits_{s \leq u} \beta_s$, *and* $\sigma = \inf\{u : \beta_u = 1\}$.

More generally, if $f_1, f_2, \ldots, f_k$ *are k functions such that:*

$$\frac{\log f_j(\lambda)}{\log \lambda} \xrightarrow[t \to \infty]{} a_j \ , \quad 1 \leq j \leq k \ ,$$

then the above convergences in law for the θ^{f_j} take place jointly, and the joint limit law is that of the vector:

$$\left(\int_0^\sigma d\gamma_s 1_{(\beta_s \leq a_j S_s)} \ ; \quad 1 \leq j \leq k \right)$$

Now, the preceding assertion about cones may be understood as a particular case of the following consequence of Theorem 7.2:
if, with the notation of Theorem 7.2 a function f satisfies: $a \geq 1$, then:

$$\frac{1}{\log t}(\theta_t - \theta_t^f) \xrightarrow[t \to \infty]{(P)} 0 \ .$$

With the help of Theorem 7.2, we are now able to present a global state-ment for asymptotic results relative to certain functionals of Brownian motion in $\mathbb{R}^3$, in the form of the following

General principle : *The limiting laws of winding numbers and, more gen-erally, of Brownian functionals in different directions of $\mathbb{R}^3$, take place jointly and independently, and, in any direction, they are given by the study in the plane, as described in the above paragraph 7.1.*

7.3 Windings of independent planar BM's around each other

The origin of the study presented in this paragraph is a question of Mitchell Berger concerning solar flares (for more details, see Berger-Roberts [5]).

Let $Z^1, Z^2, \ldots, Z^n$ be n independent planar BM's, starting from n different points $z_1, \ldots, z_n$. Then, for each $i \neq j$, we have:

$$P\left(\exists t \geq 0, Z_t^i = Z_t^j\right) = 0 \ ,$$

since $B_t^{i,j} = \frac{1}{\sqrt{2}}(Z_t^i - Z_t^j), t \geq 0$, is a planar BM starting from $\frac{1}{\sqrt{2}}(z_i - z_j) \neq 0$, and which, therefore, shall almost surely never visit 0.

Thus, we may define $(\theta_t^{i,j}, t \geq 0)$ as the winding number of $B^{i,j}$ around 0, and ask for the asymptotic law of these different winding numbers, indexed by (i,j), with $1 \leq i < j \leq n$. This is a dual situation to the situation considered in paragraph 7.1 in that, now, we consider n BM's and one point, instead of one BM and n points.

We remark that, taken all together, the processes $(B^{i,j}; 1 \leq i \leq j \leq n)$ are not independent. Nonetheless, we may prove the following result:

$$\frac{2}{\log t}\left(\theta_t^{i,j}; 1 \leq i < j \leq n\right) \xrightarrow[t \to \infty]{\text{(law)}} \left(C^{i,j}; 1 \leq i < j \leq n\right) \ , \qquad (7.7)$$

where the $C^{i,j}$'s are independent Cauchy variables, with parameter 1.

The asymptotic result (7.7) shall appear in the next paragraph as a particular case.

7.4 A unified picture of windings

The aim of this paragraph is to present a general set-up for which the studies made in paragraphs 7.1, 7.2, and 7.3, may be understood as particular cases.

Such a unification is made possible by considering: $B^1, B^2, \ldots, B^m$, m planar Brownian motions with respect to the same filtration, and which are,

moreover, linearly correlated, in the following sense:
for any $p, q \leq m$, there exists a correlation matrix $A_{p,q}$ between B^p and B^q
such that: for every $\overrightarrow{u}, \overrightarrow{v} \in \mathbb{R}^2$,

$$(\overrightarrow{u}, B_t^p)(\overrightarrow{v}, B_t^q) - (\overrightarrow{u}, A_{p,q}\overrightarrow{v})t$$

is a martingale. (Here, $(\overrightarrow{x}, \overrightarrow{y})$ denotes the scalar product in $\mathbb{R}^2$). The asymptotic result (7.7) may now be generalized as follows.

Theorem 7.3 *Let θ_t^p be the winding number of $(B_s^p, s \leq t)$ around z_0, where $B_0^p \neq z_0$, for every p.*

If, for all (p, q), $p \neq q$, the matrix $A_{p,q}$ is not an orthogonal matrix, then:

$$\frac{2}{\log t}(\theta_t^p; p \leq m) \xrightarrow[t\to\infty]{\text{(law)}} (C^p; p \leq m) \ ,$$

where the variables $(C^p; p \leq m)$ are independent Cauchy variables, with parameter 1.

The asymptotic result (7.7) appears indeed as a particular case of Theorem 7.3, since, if: $B_t^p = \frac{1}{\sqrt{2}}(Z_t^k - Z_t^\ell)$ and $B_t^q = \frac{1}{\sqrt{2}}(Z_t^k - Z_t^j)$, for $k \neq \ell \neq j$, then: $A_{p,q} = \frac{1}{2}Id$, which is not an orthogonal matrix! In other cases, $A_{p,q} = 0$.

It is natural to consider the more general situation, for which some of the matrices $A_{p,q}$ may be orthogonal. If a correlation matrix $A_{p,q}$ is orthogonal, then B^p is obtained from B^q by an orthogonal transformation and, possibly, a translation. This allows to consider the asymptotic problem in the following form: again, we may assume that none of the $A_{p,q}$'s is orthogonal, but we now have to study the winding numbers of m linearly correlated Brownian motions around n points $(z^1, \ldots, z^n)$. We write:

$$\left\{ \underline{\theta}_t^p = (\theta_t^{p,z^j}; j \leq n) \ ; \ p \leq m \right\} \ .$$

We may now state the following general result.

Theorem 7.4 *We assume that, for all $p \neq q$, $A_{p,q}$ is not orthogonal. Then,*

$$\frac{2}{\log t}(\underline{\theta}_t^p; p \leq m) \xrightarrow[t\to\infty]{\text{(law)}} (\underline{\xi}^p; p \leq m) \ ,$$

where the random vectors $(\underline{\xi}^p)_{p \leq m}$ are independent, and, for every p: $\underline{\xi}^p \overset{\text{(law)}}{=} (W_1, \ldots, W_n)$, the law of which has been described precisely in paragraph 7.1.

We now give a sketch of the main arguments in the proof of Theorem 7.4. The elementary, but nonetheless crucial fact on which the proof relies is presented in the following

Lemma 7.1 *Let G and G' be two jointly Gaussian, centered, variables in $\mathbb{R}^2$, such that: for every $u \in \mathbb{R}^2$, and every $v \in \mathbb{R}^2$,*

$$E\left[(u, G)^2\right] = |u|^2 = E\left[(u, G')^2\right] \quad , \quad \text{and} \quad E\left[(u, G)(v, G')\right] = (u, Av) \ ,$$

where A is non-orthogonal.

Then, $E\left[\dfrac{1}{|G|^p|G'|^q}\right] < \infty$, as soon as: $p < \dfrac{3}{2}, q < \dfrac{3}{2}$.

Remark: This integrability result should be compared with the fact that $E\left(\dfrac{1}{|G|^2}\right) = \infty$, which has a lot to do with the normalization of $\displaystyle\int_0^t \dfrac{ds}{|Z_s|^2}$ by $(\log t)^2$ (and not $(\log t)$, as in the Kallianpur-Robbins limit law) to obtain a limit in law.

7.5 The asymptotic distribution of the self-linking number of BM in $\mathbb{R}^3$

Gauss has defined the linking number of two closed curves in $\mathbb{R}^3$, which do not intersect each other. We should like to consider such a number for two Brownian curves, but two independent $BM's$ in $\mathbb{R}^3$ almost surely intersect each other. However, we can define some approximation to Gauss' linking number by excluding the pairs of instants (u, v) at which the two BM's are closer than $\dfrac{1}{n}$ to each other, and then let n go to infinity. It may be expected, and we shall show that this is indeed the case, that the asymptotic study shall involve some quantity related to the intersections of the two BM's.

We remark that it is also possible to define such linking number approximations for only one BM in $\mathbb{R}^3$. Thus, we consider:

$$I_n(t) \stackrel{\text{def}}{=} \int_0^t \left(dB_u, \int_0^u dB_s, \frac{B_u - B_s}{|B_u - B_s|^3}\right) 1_{(|B_u - B_s| \geq \frac{1}{n})}$$

and

$$J_n(s,t) \overset{\text{def}}{=} \int_0^s \left(dB_u, \int_0^t dB'_v, \frac{B_u - B'_v}{|B_u - B'_v|^3} \right) 1_{(|B_u-B'_v|\geq \frac{1}{n})} \, ,$$

where $(a,b,c) = a \cdot (b \times c)$ denotes the mixed product of the three vectors a, b, c in $\mathbb{R}^3$.

We have to explain the meaning given to each of the integrals:

a) in the case of J_n, there is no difficulty, since B and B' are independent,

b) in the case of I_n, we first fix u, and then:

- we either use the fact that $(B_s, s \leq u)$ is a semimartingale in the original filtration of B, enlarged by the variable B_u;

- or, we define the integral with respect to dB_s for every $x(= B_u)$, and having defined these integrals measurably in x, we replace x by B_u. Both operations give the same quantity.

We now state the asymptotic result for I_n.

Theorem 7.5 *We have:*

$$\left(B_t, \frac{1}{n} I_n(t); t \geq 0 \right) \xrightarrow[n\to\infty]{\text{(law)}} (B_t, c\beta_t; t \geq 0)$$

where (β_t) is a real-valued BM independent of B, and c is a universal constant.

To state the asymptotic result for J_n, we need to present the notion of intersection local times:
these consist in the a.s. unique family

$$(\alpha(x; s, t); x \in \mathbb{R}^3, s, t \geq 0)$$

of occupation densities, which is jointly continuous in (x, s, t), such that: for every Borel function $f : \mathbb{R}^3 \to \mathbb{R}_+$,

$$\int_0^s du \int_0^t dv \, f(B_u - B'_v) = \int_{\mathbb{R}^3} dx \, f(x)\alpha(x; s, t)$$

$(\alpha(x; du\, dv)$ is a random measure supported by $\{u, v\} : B_u - B'_v = x\})$.
The asymptotic result for J_n is the following

Theorem 7.6 *We have:*

$$\left(B_s, B'_t; \frac{1}{\sqrt{n}} J_n(s, t); s, t \geq 0\right) \xrightarrow[n\to\infty]{(\text{law})} (B_s, B'_t; c\mathbb{B}_\alpha(s, t); s, t \geq 0)$$

where c is a universal constant, and conditionally on (B, B'), the process $(\mathbb{B}_\alpha(s, t); s, t \geq 0)$ is a centered Gaussian process with covariance:

$$E\left[\mathbb{B}_\alpha(s, t)\mathbb{B}_\alpha(s', t') \mid B, B'\right] = \alpha(0; s \wedge s', t \wedge t') \ .$$

We now end up this chapter by giving a sketch of the proof of Theorem 7.5:

- in a first step, we consider, for fixed u, the sequence:

$$\theta_n(u) = \int\limits_0^u dB_s \times \frac{B_u - B_s}{|B_u - B_s|^3} 1_{\left(|B_u - B_s| \geq \frac{1}{n}\right)} \ .$$

It is then easy to show that:

$$\frac{1}{n}\theta_n(u) \xrightarrow[n\to\infty]{(\text{law})} \theta_\infty \overset{\text{def}}{=} \int\limits_0^\infty dB_s \times \frac{B_s}{|B_s|^3} 1_{(|B_s| \geq 1)} \qquad (7.8)$$

and the limit variable θ_∞ has moments of all orders, as follows from Exercise 7.1 below;

- in a second step, we remark that, for $u < v$:

$$\frac{1}{n}\left(\theta_n(u), \theta_n(v)\right) \xrightarrow[n\to\infty]{(\text{law})} (\theta_\infty, \hat{\theta}_\infty) \ , \qquad (7.9)$$

where θ_∞ and $\hat{\theta}_\infty$ are two independent copies.

To prove this result, we remark that, in the stochastic integral which defines $\theta_n(u)$, only times s which may be chosen arbitrarily close to u, and smaller than u, will make some contribution to the limit in law (7.8); then, the convergence in law (7.9) follows from the independence of the increments of B.

- in the final step, we write:

$$\frac{1}{n}I_n(t) = \gamma^{(n)}\left(\frac{1}{n^2}\int_0^t ds\,|\theta_n(s)|^2\right)\quad,$$

where $(\gamma_u^{(n)}, u \geq 0)$ is a one dimensional Brownian motion, and it is then easy to show, thanks to the results obtained in the second step that

$$\frac{1}{n^2}\int_0^t ds|\theta_n(s)|^2 \xrightarrow[n\to\infty]{L^2} c^2 t\;.$$

This convergence in L^2 follows from the convergence of the first, resp.: second, moment of the left-hand side to: $c^2 t$, resp.: $(c^2 t)^2$, as a consequence of (7.9). This allows to end up the proof of the Theorem.

Exercise 7.1 Let $(B_t, t \geq 0)$ be a 3-dimensional Brownian motion starting from 0.

1. Prove that:

$$\int_0^\infty \frac{dt}{|B_t|^4}1_{(|B_t|\geq 1)} \overset{\text{(law)}}{=} T_1^* \overset{\text{def}}{=} \inf\{u : |\beta_u| = 1\} \tag{7.10}$$

where $(\beta_u, u \geq 0)$ is a one-dimensional BM starting from 0.

Hint: Show that one may assume: $B_0 = 1$; then, prove the existence of a real-valued Brownian motion $(\gamma(u), u \geq 0)$ starting from 1, such that:

$$\frac{1}{|B_t|} = \gamma\left(\int_0^t \frac{du}{|B_u|^4}\right)\;,\quad t \geq 0\;.$$

2. Conclude that θ_∞ (defined in (7.8)) admits moments of all orders.

Hint: Apply the Burkholder-Gundy inequalities.

Exercise 7.2 (We use the same notation as in Exercise 7.1).
Prove the identity in law (7.10) as a consequence of the Ray-Knight theorem $(RK2)$, a), presented in paragraph 3.1:

$$(\ell_\infty^a(R_3), a \geq 0) \overset{\text{(law)}}{=} (R_2^2(a), a \geq 0)$$

and of the invariance by time-inversion of the law of $(R_2(a), a \geq 0)$.

Exercise 7.3 Let $(\tilde{B}_t, t \geq 0)$ be a 2-dimensional BM starting from 0, and $(\beta_t; t \geq 0)$ be a one-dimensional BM starting from 0.

1. Prove the following identities in law:

$$\left(\int_0^1 \frac{ds}{|\tilde{B}_s|} \right)^2 \stackrel{\text{(law)}}{\underset{\text{(a)}}{=}} 4 \left(\int_0^1 ds |\tilde{B}_s|^2 \right)^{-1} \stackrel{\text{(law)}}{\underset{\text{(b)}}{=}} \frac{4}{T_1^*} \stackrel{\text{(law)}}{\underset{\text{(c)}}{=}} 4 \left(\sup_{s \leq 1} |\beta_s| \right)^2$$

In particular, one has:

$$\int_0^1 \frac{ds}{|\tilde{B}_s|} \stackrel{\text{(law)}}{=} 2 \sup_{s \leq 1} |\beta_s| \tag{7.11}$$

Hints: To prove (a), represent $(|\tilde{B}_s|^2, s \geq 0)$ as another 2-dimensional Bessel process, time-changed; to prove (b), use, e.g., the Ray-Knight theorem on Brownian local times; to prove (c), use the scaling property.

2. Define $S = \inf \left\{ u : \int_0^u \frac{ds}{|\tilde{B}_s|} > 1 \right\}$. Deduce from (7.11) that:

$$S \stackrel{\text{(law)}}{=} \frac{T_1^*}{4}$$

and, consequently:

$$E \left[\exp \left(-\frac{\lambda^2}{2} S \right) \right] = \left(\text{ch} \left(\frac{\lambda}{2} \right) \right)^{-1} \tag{7.12}$$

Comments on Chapter 7

The proofs of the results presented in paragraph 7.1 are found in Pitman-Yor ([75], [76]); those in paragraph 7.2 are found in Le Gall-Yor [61], and the results in paragraphs 7.3 and 7.4 are taken from Yor [100].

The asymptotic study of windings for random walks has been made by Belisle [3] (see also Belisle-Faraway [4]); there are also many publications on this topic in the physics literature (see, e.g., Rudnick-Hu [82]).

The proof of Theorem 7.5 constitutes a good example that the asymptotic study of some double integrals with respect to BM may, in a number of cases, be reduced to a careful study of simple integrals (see, e.g., the reference to Stroock-Varadhan-Papanicolaou in Chapter XIII of Revuz-Yor [81]).

Chapter 8

Some extensions of Paul Lévy's arc sine law for BM

In his 1939 paper: "Sur certains processus stochastiques homogènes", Paul Lévy [64] proves that both Brownian variables:

$$A^+ \stackrel{\text{def}}{=} \int_0^1 ds \, 1_{(B_s>0)} \quad \text{and} \quad g = \sup\{t < 1 : B_t = 0\}$$

are arc-sine distributed.

Over the years, these results have been extended in many directions; for a review of extensions developed up to 1988, see Bingham-Doney [20].

In this Chapter, we present further results, which extend Lévy's computation in the three following directions, in which $(B_t, t \geq 0)$ is replaced respectively by:

i) a symmetrized Bessel process with dimension $0 < \delta < 2$,

ii) a Walsh Brownian motion, that is a process $(X_t, t \geq 0)$ in the plane which takes values in a finite number of rays ($\equiv$ half-lines), all meeting at 0, and such that $(X_t, t \geq 0)$ behaves, while away from 0, as a Brownian motion, and, when it meets 0, chooses a ray with equal probability,

iii) a singularly perturbed reflecting Brownian motion, that is $(|B_t| - \mu \ell_t, t \geq 0)$ where $(\ell_t, t \geq 0)$ is the local time of $(B_t, t \geq 0)$ at 0.

A posterior justification of these extensions may be that the results which one obtains in each of these directions are particularly simple, this being due partly to the fact that, for each of the models, the strong Markov property and the scaling property are available; for example, in the set-up of (iii), we rely upon the strong Markov property of the 2-dimensional process: $\{|B_t|, \ell_t; t \geq 0\}$.

More importantly, these three models may be considered as testing grounds for the use and development of the main methods which have been successful in recent years in reproving Lévy's arc sine law, that is, essentially: excursion theory and stochastic calculus (more precisely, Tanaka's formula).

Finally, one remarkable feature in this study needs to be underlined: although the local time at 0 of, say, Brownian motion, does not appear a priori in the problem studied here, that is: determining the law of A^+, in fact, it plays an essential role, and a main purpose of this chapter is to clarify this role.

8.1 Some notation

Throughout this chapter, we shall use the following notation:
Z_a, resp.: $Z_{a,b}$, denotes a gamma variable with parameter a, resp.: a beta variable with parameters (a, b), so that

$$P(Z_a \in dt) = \frac{dt\ t^{a-1}e^{-t}}{\Gamma(a)}(t > 0)$$

and

$$P(Z_{a,b} \in dt) = \frac{dt\ t^{a-1}(1-t)^{b-1}}{B(a,b)} \quad (0 < t < 1)$$

We recall the well-known algebraic relations between the laws of the beta and gamma variables:

$$Z_a \stackrel{\text{(law)}}{=} Z_{a,b}Z_{a+b} \quad \text{and} \quad Z_{a,b+c} \stackrel{\text{(law)}}{=} Z_{a,b}Z_{a+b,c} \ ,$$

where, in both identities in law, the right-hand sides feature independent r.v.'s. We shall also use the notation $T_{(\alpha)}$, with $0 < \alpha < 1$, to denote a one-sided stable (α) random variable, the law of which may be characterized by:

$$E\left[\exp(-\lambda T_{(\alpha)})\right] = \exp(-\lambda^\alpha) \ , \quad \lambda \geq 0 \ .$$

(It may be worth noting that $2T_{(1/2)}$, and not $T_{(1/2)}$, is distributed as the first hitting time of 1 by a one-dimensional BM starting from 0).

8.2 A list of results

(8.2.1)   As was already recalled, Lévy (1939) proved that A^+ and g are arc sine distributed, that is: they have the same law as $\dfrac{N^2}{N^2 + N'^2}$, where N and N' are two centered, independent Gaussian variables with variance 1, or, since: $N^2 \overset{\text{(law)}}{=} \dfrac{1}{2T_{(1/2)}}$, we see that A^+ and g are distributed as:

$$\frac{T_{(1/2)}}{T_{(1/2)} + T'_{(1/2)}} \tag{8.1}$$

where $T_{(1/2)}$ and $T'_{(1/2)}$ are two independent copies. In fact, in the next paragraph, we shall present some proofs which exhibit A^+ in the form (8.1).

For the moment, here is a quick proof that g is arc-sine distributed:
let $u \le 1$; then: $(g < u) = (d_u > 1)$,
where: $d_u = \inf\{t \ge u; B_t = 0\}$
$$\equiv u + \inf\{v > 0 : B_{v+u} - B_u = -B_u\}$$
$$\overset{\text{(law)}}{=} u + B_u^2 \sigma \overset{\text{(law)}}{=} u(1 + B_1^2 \sigma),$$
with: $\sigma = \inf\{t : \beta_t = 1\}$, and β is a BM, independent of B_u. Hence, we have shown:
$$g \overset{\text{(law)}}{=} 1 + B_1^2 \sigma \overset{\text{(law)}}{=} 1 + \frac{B_1^2}{\beta_1^2} \overset{\text{(law)}}{=} 1 + \frac{N^2}{N'^2}, \text{ which gives the result.}$$

(8.2.2)   If we replace Brownian motion by a symmetrized Bessel process of dimension $0 < \delta = 2(1 - \alpha) < 2$, then the quantities $A^+_{(\alpha)}$ and $g_{(\alpha)}$, the meaning of which is self-evident, no longer have a common distribution if $\alpha \ne \frac{1}{2}$. In fact, Dynkin [38] showed that: $g_{(\alpha)} \overset{\text{(law)}}{=} Z_{\alpha, 1-\alpha}$, whereas Barlow-Pitman-Yor [2] proved that:

$$A^+_{(\alpha)} \overset{\text{(law)}}{=} \frac{T_{(\alpha)}}{T_{(\alpha)} + T'_{(\alpha)}}, \tag{8.2}$$

where $T_{(\alpha)}$ and $T'_{(\alpha)}$ are two independent copies.

(8.2.3)   In [2], it was also shown that Lévy's result for A^+ admits the following multivariate extension: if we consider (as described informally in the introduction to this chapter) a Walsh Brownian motion $(Z_s, s \ge 0)$ living on n rays $(I_i; 1 \le i \le n)$, and we denote:

$$A^{(i)} = \int_0^1 ds\, 1_{(Z_s \in I_i)} \;,$$

then:

$$\left(A^{(1)}, \ldots, A^{(n)}\right) \overset{\text{(law)}}{=} \left(\frac{T^{(i)}}{\sum\limits_{j=1}^n T^{(j)}} \;; 1 \le i \le n\right) \tag{8.3}$$

where $(T^{(i)}; i \le i \le n)$ are n independent one-sided stable $\left(\frac{1}{2}\right)$ random variables. Furthermore, it is possible to give a common extension of (8.2) and (8.3), by considering a process $(Z_s, s \ge 0)$ which, on each of the rays, behaves like a Bessel process with dimension $\delta = 2(1 - \alpha)$, and when arriving at 0, chooses its ray with equal probability. Then, using a self-evident notation, we have:

$$\left(A_{(\alpha)}^{(1)}, \ldots, A_{(\alpha)}^{(n)}\right) \overset{\text{(law)}}{=} \left(\frac{T_{(\alpha)}^{(i)}}{\sum\limits_{j=1}^n T_{(\alpha)}^{(j)}} \;; 1 \le i \le n\right) \;. \tag{8.4}$$

(8.2.4)　　However, in this chapter, we shall be more concerned with yet another family of extensions of Lévy's results, which have been obtained by F. Petit in her thesis [70].

Theorem 8.1 *For any $\mu > 0$, we have*

$$\int_0^1 ds\, 1_{(|B_s| \le \mu \ell_s)} \overset{\text{(law)}}{=} Z_{\frac{1}{2}, \frac{1}{2\mu}} \;, \tag{8.5}$$

and

$$\int_0^g ds\, 1_{(|B_s| \le \mu \ell_s)} \overset{\text{(law)}}{=} Z_{\frac{1}{2}, \frac{1}{2} + \frac{1}{2\mu}} \;. \tag{8.6}$$

In the sequel, we shall refer to the identities in law (8.5) and (8.6) as to F. Petit's first, resp. second result.

With the help of Lévy's identity in law:

$$(S_t - B_t, S_t; t \ge 0) \overset{\text{(law)}}{=} (|B_t|, \ell_t; t \ge 0) \;,$$

and Pitman's theorem ([71]):

$$(2S_t - B_t, S_t; t \geq 0) \overset{(\text{law})}{=} (R_t, J_t; t \geq 0) \ ,$$

where $(R_t, t \geq 0)$ is a 3-dimensional Bessel process starting from 0, and $J_t = \inf_{s \geq t} R_s$, we may translate, for example, (8.5) in the following terms:

$$\int_0^1 ds \ 1_{(B_s \geq (1-\mu)S_s)} \overset{(\text{law})}{=} \int_0^1 ds \ 1_{(R_s \leq (1+\mu)J_s)} \overset{(\text{law})}{=} Z_{\frac{1}{2}, \frac{1}{2\mu}} \qquad (8.7)$$

which shows, in particular, that for $\mu = 1$, the result agrees with Lévy's arc sine law.

Using the representation of the standard Brownian bridge $(b(u), u \leq 1)$ as:

$$\left(\frac{1}{\sqrt{g}} B_{gu} \ , \quad u \leq 1 \right)$$

and the independence of this process from g, we may deduce from (8.6) the following

Corollary 8.1.1   *Let $(b(u), u \leq 1)$ be a standard Brownian bridge, and $(\lambda_u, u \leq 1)$ be its local time at 0. Then, we have*

$$\int_0^1 ds \ 1_{(|b(s)| \leq \mu \lambda_s)} \overset{(\text{law})}{=} Z_{1, \frac{1}{2\mu}} \overset{(\text{law})}{=} 1 - U^{\frac{1}{2\mu}} \ , \qquad (8.8)$$

where U is uniformly distributed on $[0, 1]$.

In particular, in the case $\mu = \frac{1}{2}$, we obtain:

$$\int_0^1 ds \ 1_{(|b(s)| \leq \frac{1}{2}\lambda_s)} \overset{(\text{law})}{=} \int_0^1 ds \ 1_{(|b(s)| + \frac{1}{2}\lambda(s) \leq \frac{1}{2}\lambda(1))} \overset{(\text{law})}{=} U \qquad (8.9)$$

Using now the following identity in law (8.10) between the Brownian Bridge $(b(u), u \leq 1)$ and the Brownian meander: $\left(m(u) \equiv \frac{1}{\sqrt{1-g}} |B_{g+u(1-g)}|, u \leq 1 \right)$:

$$\left(m(s), j(s) \equiv \inf_{s \leq u \leq 1} m(u); s \leq 1 \right) \overset{(\text{law})}{=} (|b(s)| + \lambda(s), \lambda(s); s \leq 1) \quad (8.10)$$

which is found in Biane-Yor [18], and Bertoin-Pitman [11], we obtain the

Corollary 8.1.2 *Let $(m(s), s \leq 1)$ denote the Brownian meander. Then we have:*

$$\int_0^1 ds 1_{(m(s) + (\mu - 1)j_s \leq \mu m_1)} \overset{(\text{law})}{=} Z_{1, \frac{1}{2\mu}}$$

In particular, we obtain, by taking $\mu = \frac{1}{2}$ and $\mu = 1$:

$$\int_0^1 ds \, 1_{\left(m(s) - \frac{1}{2}j(s) \leq \frac{1}{2}m(1) \right)} \overset{(\text{law})}{=} U \ ,$$

and

$$P \left\{ \int_0^1 ds \, 1_{(m(s) \geq m(1))} \in dt \right\} = \frac{dt}{2\sqrt{t}} \quad .$$

PROOF: Together with the identity in law (8.10), we use the symmetry of the law of the Brownian bridge by time reversal, i.e.:

$$(b(u), u \leq 1) \overset{(\text{law})}{=} (b(1 - u), u \leq 1) \ .$$

We then obtain:

$$\int_0^1 ds \, 1_{(|b(s)| \leq \mu \lambda_s)} \overset{(\text{law})}{=} \int_0^1 ds \, 1_{(|b(s)| \leq \mu(\lambda_1 - \lambda_s))} \overset{(\text{law})}{=} \int_0^1 ds \, 1_{(m(s) + (\mu - 1)j_s < \mu m_1)} \ ,$$

and the desired results follow from Corollary 8.1.1. $\square$

8.3 A discussion of methods - Some proofs

(8.3.1) We first show how to prove

$$A_1^+ \equiv \int_0^1 ds \, 1_{(B_s>0)} \stackrel{(law)}{=} Z_{\frac{1}{2},\frac{1}{2}}$$

by using jointly the scaling property of Brownian motion, and excursion theory. Set $A_t^+ = \int_0^t ds \, 1_{(B_s>0)}$ and $A_t^- = \int_0^t ds \, 1_{(B_s<0)}$ $(t \geq 0)$.

We have, for every t, and u: $(A_t^+ > u) = (t > \alpha_u^+)$,
where $\alpha_u^+ \stackrel{def}{=} \inf\{s; A_s^+ > u\}$. We now deduce, by scaling, that:

$$A_1^+ \stackrel{(law)}{=} \frac{1}{\alpha_1^+} \tag{8.11}$$

From the trivial identity: $t = A_t^+ + A_t^-$, it follows: $\alpha_u^+ = u + A_{\alpha_u^+}^-$;
then, we write: $A_{\alpha_u^+}^- = A_{\tau(\ell_{\alpha_u^+})}^-$, with $\tau(s) = \inf\{v; \ell_v > s\}$.

Now, it is a consequence of excursion theory that the two processes $(A_{\tau(t)}^+, t \geq 0)$ and $(A_{\tau(t)}^-, t \geq 0)$ are independent; hence, the two processes $(A_{\tau(t)}^-, t \geq 0)$ and $(\ell_{\alpha_u^+}, u \geq 0)$ are independent; consequently, we now deduce from the previous equalities that, for fixed u:

$$\alpha_u^+ \stackrel{(law)}{=} u + (\ell_{\alpha_u^+})^2 A_{\tau(1)}^-$$

$$\stackrel{(law)}{=} u \left(1 + \frac{A_{\tau(1)}^-}{A_{\tau(1)}^+}\right) \quad , \quad \text{again by scaling} \tag{8.12}$$

Putting together (8.11) and (8.12), we obtain:

$$A_1^+ \stackrel{(law)}{=} \frac{A_{\tau(1)}^+}{A_{\tau(1)}^+ + A_{\tau(1)}^-} \quad .$$

Now, from $(RK1)$ in paragraph 3.1, we know that:

$$\left(A_{\tau(1)}^+, A_{\tau(1)}^-\right) \stackrel{(law)}{=} \frac{1}{2}(T_{(1/2)}, T'_{(1/2)}) \quad ,$$

from which we obtain the representation (8.1) for A_1^+, hence:

$$A_1^+ \stackrel{(law)}{=} Z_{1/2,1/2}$$

(8.3.2) It may also be interesting to avoid using the scaling property, and only depend on the excursion theory arguments, so that the method may be used for diffusions which do not possess the scaling property; see some extensions of the arc-sine law to real-valued diffusions by A. Truman and D. Williams ([86], [87]).

Recall that, from the master formulae of excursion theory (see Proposition 3.2), we have, for every continuous, positive, additive functional $(A_t, t \geq 0)$:

$$E_0 \left[\exp(-\lambda A_{S_\theta}) \right]$$
$$= \frac{\theta^2}{2} \int_0^\infty ds E_0 \left[\exp \left(-\lambda A_{\tau_s} - \frac{\theta^2}{2} \tau_s \right) \right] \int_{-\infty}^\infty da \, E_a \left[\exp \left(-\lambda A_{T_0} - \frac{\theta^2}{2} T_0 \right) \right] .$$

Applying this formula to $A = A^+$, we remark that:

- on one hand,

$$E_0 \left[\exp \left(-\lambda A_{\tau_s}^+ - \frac{\theta^2}{2} \tau_s \right) \right] = E_0 \left[\exp - \left(\lambda + \frac{\theta^2}{2} \right) A_{\tau_s}^+ \right] E_0 \left[\exp \left(-\frac{\theta^2}{2} A_{\tau_s}^- \right) \right]$$

$$= \exp \left(-\frac{s}{2} \sqrt{2\lambda + \theta^2} \right) \exp \left(-\frac{s\theta}{2} \right) ;$$

- on the other hand:

$$E_a \left[\exp \left(-\lambda A_{T_0} - \frac{\theta^2}{2} T_0 \right) \right]$$

$$= \begin{cases} E_a \left[\exp - \left(\lambda + \frac{\theta^2}{2} \right) T_0 \right] = \exp -a\sqrt{2\lambda + \theta^2} , & \text{if } a > 0; \\ E_a \left[\exp - \left(\frac{\theta^2}{2} T_0 \right) \right] = \exp(-|a|\theta) , & \text{if } a < 0; \end{cases}$$

Consequently, we obtain:

$$E_0 \left[\exp(-\lambda A_{S_\theta}) \right] = \frac{\theta^2}{\left(\sqrt{2\lambda + \theta^2} + \theta \right)} \left(\frac{1}{\sqrt{2\lambda + \theta^2}} + \frac{1}{\theta} \right) ,$$

from which, at least in theory, one is able to deduce, by inversion of the Laplace transform in θ, that:

$$A_t^+ \overset{\text{(law)}}{=} t \, Z_{\frac{1}{2}, \frac{1}{2}} .$$

Remark: This approach is the excursion theory variant of the Feynman-Kac approach; see, for example, Itô-Mc Kean ([50], p. 57–58).

(8.3.3) It is not difficult, with the help of the master formulae of excursion theory (see Proposition 3.2), to enlarge the scope of the above method and, using the scaling property again, Barlow-Pitman-Yor [2] arrived to the following identity in law:
for every $t > 0$ and $s > 0$,

$$\frac{1}{\ell_t^2}(A_t^+, A_t^-) \stackrel{(law)}{=} \frac{1}{s^2}(A_{\tau_s}^+, A_{\tau_s}^-)$$

(by scaling, the left-hand side is equal in law to: $\frac{1}{\ell_{S_\theta}^2}(A_{S_\theta}^+, A_{S_\theta}^-)$, for every $\theta > 0$, which enables to use the master formula of excursion theory).

Hence, we have:

$$\frac{1}{\ell_t^2}(A_t^+, A_t^-) \stackrel{(law)}{=} \frac{1}{4}\left(T_{(\frac{1}{2})}, T'_{(\frac{1}{2})}\right) \ ,$$

which implies (8.1): $A_1^+ \stackrel{(law)}{=} \dfrac{T_{(\frac{1}{2})}}{T_{(\frac{1}{2})} + T'_{(\frac{1}{2})}}$, i.e.: A_1^+ is arc-sine distributed.

Pitman-Yor [77] give a more complete explanation of the fact that:

$$\frac{1}{\ell_T^2}(A_T^+, A_T^-)$$

has a distribution which does not depend on T, for a certain class of random variables; besides the case $T = t$, another interesting example is:

$$T \equiv \alpha_t^+ \equiv \inf\{u : A_u^+ > t\}.$$

By analogy, F. Petit's original results (Theorem 8.1 above), together with the arithmetic of beta-gamma laws led us to think that the four pairs of random variables:

$$(8.13) \qquad \frac{1}{(\ell_t^\mu)^2}\left(A_t^{\mu,-}, A_t^{\mu,+}\right) \ ; \quad (8.14) \qquad \frac{1}{t^2}\left(A_{\tau_t^\mu}^{\mu,-}, A_{\tau_t^\mu}^{\mu,+}\right) \ ;$$

$$(8.15) \qquad \frac{1}{(\ell_{\alpha_s^{\mu,-}}^\mu)^2}\left(s, A_{\alpha_s^{\mu,-}}^{\mu,+}\right) \ ; \quad (8.16) \qquad \frac{1}{8}\left(\frac{1}{Z_{\frac{1}{2\mu}}}, \frac{1}{Z_{\frac{1}{2}}}\right)$$

may have the same distribution. This is indeed true, as we shall see partly in the sequel. (Here, and in the following, $(\ell_t^\mu, t \geq 0)$ denotes the (semi-martingale) local time at 0 of $(|B_t| - \mu\ell_t, t \geq 0), (\tau_t^\mu, t \geq 0)$ is the inverse of $(\ell_t^\mu, t \geq 0)$, and $(\alpha_t^{\mu,-}, t \geq 0)$ is the inverse of $(A_t^{\mu,-}, t \geq 0))$. It may be worth, to give a better understanding of the identity in law between (8.15) and (8.16), to present this identity in the following equivalent way:

Theorem 8.2 *1) The identity in law*

$$\frac{1}{8} \left(\frac{\ell_{\alpha_1}^\mu}{\alpha_1 - 1}, (\ell_{\alpha_1}^\mu)^2 \right) \overset{\text{(law)}}{=} \left(Z_{\frac{1}{2}}, Z_{\frac{1}{2\mu}} \right) \tag{8.17}$$

holds. (Here, we have written, for clarity, α_1 for $\alpha_1^{\mu,-}$).

2) Consequently, we have:

$$A_1^{\mu,-} \overset{\text{(law)}}{=} \frac{1}{\alpha_1^{\mu,-}} \overset{\text{(law)}}{=} \frac{1}{1 + \frac{Z_{1/2\mu}}{Z_{1/2}}} \overset{\text{(law)}}{=} Z_{\frac{1}{2}, \frac{1}{2\mu}} \ .$$

Comment: The second statement of this Theorem is deduced immediately from the first one, using the scaling property; it gives an explanation of F. Petit's first result.

(8.3.4)   To end up our discussion of methods, we now mention that Knight's theorem about continuous orthogonal martingales may replace the excursion argument to prove the independence of the processes $(A_{\tau_t}^+, t \geq 0)$ and $(A_{\tau_t}^-, t \geq 0)$. To see this, we remark that Tanaka's formula and Knight's theorem, used jointly, imply:

$$B_t^+ = -\beta_{A_t^+}^{(+)} + \frac{1}{2}\ell_t \quad \text{and} \quad B_t^- = -\beta_{A_t^-}^{(-)} + \frac{1}{2}\ell_t \ ,$$

with: $\beta^{(+)}$ and $\beta^{(-)}$ two independent BM's, and:

$$A_{\tau_t}^\pm = \inf \left\{ u : \beta_u^{(\pm)} = \frac{1}{2}t \right\} \ .$$

In the last paragraph 8.5 of this Chapter, we shall see how to modify this argument when (B_t) is replaced by $(|B_t| - \mu\ell_t, t \geq 0)$.

8.4 An excursion theory approach to F. Petit's results

(8.4.1)   As we remarked in paragraph 8.3, F. Petit's first result:

$$A_1^{\mu,-} \stackrel{\text{def}}{=} \int_0^1 ds 1_{(|B_s| \leq \mu \ell_s)} \stackrel{\text{(law)}}{=} Z_{1/2,1/2\mu} \tag{8.5}$$

is equivalent to (see formula (8.11)):

$$\frac{1}{\alpha_1^{\mu,-}} \stackrel{\text{(law)}}{=} Z_{1/2,1/2\mu} \tag{8.18}$$

To simplify notation, we shall simply write, in the sequel, A, Z, and α for, respectively: $A_1^{\mu,-}$, $Z_{1/2,1/2\mu}$ and $\alpha_1^{\mu,-}$.

To prove (8.5) or equivalently (8.18), we shall compute the following quantity:

$$E\left[\exp\left(-\frac{\lambda^2}{2}A_{S_\theta}\right)\varphi\left(|B_{S_\theta}|, \ell_{S_\theta}\right)\right] \equiv \frac{\theta^2}{2}\int_0^\infty dt\, e^{-\frac{\theta^2 t}{2}} E\left[e^{-\frac{\lambda^2}{2}A_t}\varphi\left(|B_t|, \ell_t\right)\right] \tag{8.19}$$

where $\varphi : \mathbb{R}_+ \times \mathbb{R}_+ \to \mathbb{R}_+$ is a Borel function, and S_θ denotes an independent exponential time with parameter $\frac{\theta^2}{2}$.

We are able to compute this quantity thanks to the extensions of the RK theorems obtained in Chapter 3 (to be more precise, see Theorem 3.4, and the computations made in subparagraph (3.3.2)), and therefore, in some sense, we may envision F. Petit's results as consequences of the extended RK theorems. However, before we embark precisely in this computation, it may be of some interest to play a little more with the scaling property; this leads us, at no cost, to the following reinforcement of (8.5).

Theorem 8.3 Let $Z \stackrel{\text{(law)}}{=} Z_{1/2,1/2\mu}$. Then, we have the following

1) $P\left(|B_1| \leq \mu \ell_1\right) = E(Z) = \frac{\mu}{1+\mu}$

2) Conditioned on the set $\Gamma_\mu \equiv (|B_1| \leq \mu \ell_1)$, the variable A_1 is distributed as $Z_{3/2,1/2\mu}$.

3) Conditioned on Γ_μ^c, A_1 is distributed as: $Z_{1/2,1+1/2\mu}$.

4) A_1 is distributed as Z.

These four statements may also be presented in the equivalent form:

$$A_1 \overset{\text{(law)}}{=} Z \quad \text{and} \quad P(\Gamma_\mu \mid A_1 = a) = a \ .$$

Remark: In fact, using the identity in law between (8.13) and (8.15), it is not difficult to prove the more general identity:

$$P\left(\Gamma_\mu \mid A_1 = a, \ell_1^\mu\right) = a$$

PROOF OF THE THEOREM:

i) These four statements may be deduced in an elementary way from the two identities:
$$E\left[\Gamma_\mu; \exp(-\alpha A_1)\right] = E\left[Z \exp(-\alpha Z)\right] \tag{8.20}$$
and
$$E\left[\exp(-\alpha A_1)\right] = E\left[\exp(-\alpha Z)\right] \tag{8.21}$$
which are valid for every $\alpha \geq 0$.

The identity (8.21) is rephrasing F. Petit's result (8.5), so that, for the moment, it remains to prove (8.20).

ii) For this purpose, we shall consider the quantity (8.19), in which we take: $\varphi(x, \ell) = 1_{(x \leq \mu\ell)}$. We then obtain:

$$E\left[\exp\left(-\frac{\lambda^2}{2}A_{S_\theta}\right)1_{\left(|B_{S_\theta}|\le\mu\ell_{S_\theta}\right)}\right]$$

$$=\frac{\theta^2}{2}E\left[\int_0^\infty dA_t \exp-\frac{1}{2}(\theta^2 t+\lambda^2 A_t)\right]$$

$$=\frac{\theta^2}{2}E\left[\int_0^\infty ds \exp-\frac{1}{2}(\theta^2\alpha_s+\lambda^2 s)\right] \quad , \quad \text{by time changing}$$

$$=\frac{\theta^2}{2}E\left[\int_0^\infty ds \exp-\frac{1}{2}(\theta^2 s\alpha_1+\lambda^2 s)\right] \quad , \quad \text{by scaling}$$

$$=\frac{\theta^2}{2}E\left[\int_0^\infty ds \exp-\frac{1}{2}\left(\frac{\theta^2 s}{A_1}+\lambda^2 s\right)\right] \quad , \quad \text{by scaling again}$$

$$=\frac{\theta^2}{2}E\left[A_1\int_0^\infty du \exp-\frac{1}{2}(\theta^2 u+\lambda^2 uA_1)\right] , \text{by change of variables: } s=A_1 u.$$

$$=E\left[A_1\exp\left(-\frac{\lambda^2}{2}S_\theta A_1\right)\right] \quad .$$

Comparing now the two extreme terms of this sequence of equalities, we obtain, by using the scaling property once again:

$$E\left[\exp\left(-\frac{\lambda^2}{2}S_\theta A_1\right)1_{(|B_1|\le\mu\ell_1)}\right]=E\left[A_1\exp\left(-\frac{\lambda^2}{2}S_\theta A_1\right)\right] \quad (8.22)$$

Since this relation is true for every $\theta>0$, we have obtained, thanks to the injectivity of the Laplace transform, that, for every $\alpha\ge 0$:

$$E\left[\exp(-\alpha A_1)1_{(|B_1|\le\mu\ell_1)}\right]=E\left[A_1\exp(-\alpha A_1)\right] \quad , \quad (8.23)$$

which proves (8.20), assuming F. Petit's result (8.5) □

Remarks:

1) The first statement of the theorem, namely:

$$P(|B_1| \leq \mu\ell_1) = \frac{\mu}{1+\mu}:$$

is an elementary consequence of the fact that, conditionally on $R \overset{\text{def}}{=} |B_1| + \ell_1$, $|B_1|$ is uniformly distributed on $[0, R]$; hence, if U denotes a uniform r.v. on $[0,1]$, which is independent from R, we have:

$$P(|B_1| \leq \mu\ell_1) = P(RU \leq \mu R(1 - U)) = P(U \leq \mu(1 - U)) = \frac{\mu}{1+\mu} \quad .$$

2) Perhaps we should emphasize the fact that the obtention of (8.23) in part (ii) of the proof of the Theorem was done with the only use of the scaling property; in particular, for this result, no knowledge of F. Petit's results is needed whatsoever.

(8.4.2)   We now engage properly into the proof of (8.5), by computing explicitly the quantity

$$\gamma_{\theta,\lambda} \overset{\text{def}}{=} E\left[\exp\left(-\frac{\lambda^2}{2}A_{S_\theta}\right)\mathbf{1}_{\left(|B_{S_\theta}| \leq \mu\ell_{S_\theta}\right)}\right] \quad . \tag{8.24}$$

We first recall that, as a consequence of the master formulae of excursion theory, we have, if we write:

$$A_t = A_t' + A_t'' \quad , \qquad \text{where:} \quad A_t' = A_{g_t} \quad \text{and} \quad A_t'' = A_t - A_{g_t} \quad ,$$

$$E\left[\exp\left(-\frac{\lambda^2}{2}A_{S_\theta}'\right)\bigg|\ell_{S_\theta} = s, B_{S_\theta} = a\right] = E_0\left[\exp-\left(\frac{\lambda^2}{2}A_{T_s} + \frac{\theta^2}{2}T_s\right)\right]e^{\theta s} \tag{8.25}$$

and

$$E\left[\exp\left(-\frac{\lambda^2}{2}A_{S_\theta}''\right)\bigg|\ell_{S_\theta} = s, B_{S_\theta} = a\right] = E_a\left[\exp-\left(\frac{\lambda^2}{2}A_{T_0}^s + \frac{\theta^2}{2}T_0\right)\right]e^{\theta|a|} \tag{8.26}$$

Moreover, from the extensions of the RK theorems obtained in Chapter 3 (see Theorem 3.4, and the computations made in subparagraph (3.3.2)), we have, by denoting: $b = \mu s$, $\nu = \sqrt{\lambda^2 + \theta^2}$, and $\xi = \frac{\theta}{\nu}$:

$$E_0\left[\exp-\left(\frac{\lambda^2}{2}A_{T_s} + \frac{\theta^2}{2}T_s\right)\right] = (\text{ch}(\nu b) + \xi\text{sh}(\nu b))^{-1/\mu} \tag{8.27}$$

$$E_a\left[\exp-\left(\frac{\lambda^2}{2}A_{T_0}^s + \frac{\theta^2}{2}T_0\right)\right] = \frac{\text{ch}(\nu(b-a)) + \xi\text{sh}(\nu(b-a))}{\text{ch}(\nu b) + \xi\text{sh}(\nu b)}, \text{ for } 0 \leq a \leq b \tag{8.28}$$

Consequently, using moreover the fact that ℓ_{S_θ} and B_{S_θ} are independent and distributed as:

$$P(\ell_{S_\theta} \in ds) = \theta e^{-\theta s} ds \quad \text{and} \quad P(B_{S_\theta} \in da) = \frac{\theta}{2} e^{-\theta |a|} da \ ,$$

we obtain:

$$\gamma_{\theta,\lambda} = \frac{\theta^2}{\mu} \int_0^\infty db \int_0^b da \frac{\text{ch}(\nu(b-a)) + \xi \text{sh}(\nu(b-a))}{(\text{ch}(\nu b) + \xi \text{sh}(\nu b))^{1+\frac{1}{\mu}}}$$

Integrating with respect to da, and making the change of variables $x = \nu b$, we obtain:

$$\gamma_{\theta,\lambda} = \frac{\xi^2}{\mu} \int_0^\infty dx \frac{\text{sh} x + \xi(\text{ch } x - 1)}{(\text{ch } x + \xi \text{ sh} x)^{1+\frac{1}{\mu}}} = \xi^2 \left(1 - \frac{\xi}{\mu} \int_0^\infty \frac{dx}{(\text{ch } x + \xi \text{ sh} x)^{1+\frac{1}{\mu}}} \right) .$$

On the other hand, we know, from (8.22), that the quantity $\gamma_{\theta,\lambda}$ is equal to:

$$E\left[A_1 \exp\left(-\frac{\lambda^2}{2} S_\theta A_1 \right) \right] = \xi^2 E\left[\frac{A_1}{A_1 + \xi^2(1 - A_1)} \right]$$

(the expression on the right-hand side is obtained after some elementary change of variables). Hence, the above computations have led us to the formula:

$$E\left[\frac{A_1}{A_1 + \xi^2(1 - A_1)} \right] = 1 - \frac{\xi}{\mu} \int_0^\infty \frac{dx}{(\text{ch } x + \xi \text{ sh} x)^{1+\frac{1}{\mu}}} \ ,$$

or, equivalently:

$$E\left[\frac{1 - A_1}{A_1 + \xi^2(1 - A_1)} \right] = \frac{1}{\xi \mu} \int_0^\infty \frac{dx}{(\text{ch } x + \xi \text{ sh} x)^{1+\frac{1}{\mu}}} \tag{8.29}$$

We now make the change of variables: $u = (\text{th} x)^2$, to obtain:

$$h(\xi) \stackrel{\text{def}}{=} E\left[\frac{\xi(1 - A_1)}{A_1 + \xi^2(1 - A_1)} \right] = \frac{1}{2\mu} \int_0^1 du (1 - u)^{\frac{1}{2\mu} - \frac{1}{2}} u^{-\frac{1}{2}} (1 + \xi\sqrt{u})^{-(1+\frac{1}{\mu})}$$

We define $r = \frac{1}{2} + \frac{1}{2\mu}$, and we use the elementary identity:

$$\frac{1}{(1+x)^p} = E\left[\exp(-x Z_p) \right]$$

to obtain:

$$h(\xi) = \frac{1}{2\mu} \int\limits_0^1 du\ u^{-\frac{1}{2}} (1-u)^{r-1} E\left[\exp(-\xi \sqrt{u} Z_{2r})\right] = c_\mu E\left[\exp -\xi Z_{2r} \sqrt{Z_{\frac{1}{2},r}}\right]$$

$$(8.30)$$

where c_μ is a constant depending only on μ, and Z_{2r} and $Z_{\frac{1}{2},r}$ are independent. The following lemma shall play a crucial role in the sequel of the proof.

Lemma 8.1 *The following identities in law hold:*

$$Z_{2r}^2 \overset{(\text{law})}{=} 4 Z_{r+\frac{1}{2}} Z_r \tag{8.31}$$

$$Z_{2r}\sqrt{Z_{\frac{1}{2},r}} \overset{(\text{law})}{=} 2\sqrt{Z_{\frac{1}{2}} Z_r} \overset{(\text{law})}{=} |N|\sqrt{2 Z_r}\ . \tag{8.32}$$

As usual, in all these identities in law, the pairs of random variables featured in the different products are independent.

PROOF OF THE LEMMA:

1) The duplication formula for the gamma function:

$$\sqrt{\pi} \Gamma(2z) = 2^{2z-1} \Gamma\left(z + \frac{1}{2}\right) \Gamma(z)$$

implies that, since for any $k > 0$, we have:

$$E[Z_p^k] = \frac{\Gamma(p+k)}{\Gamma(p)}\ ,$$

then:

$$E[Z_{2r}^{2k}] = 4^k E[Z_{r+\frac{1}{2}}^k] E[Z_r^k]\ .$$

2) The first identity in law in (8.32) follows from (8.31), and the fact that:
$$Z_{\frac{1}{2},r} Z_{\frac{1}{2}+r} \overset{(\text{law})}{=} Z_{1/2},$$ and the second identity in law is immediate since:

$$|N| \overset{(\text{law})}{=} \sqrt{2 Z_{\frac{1}{2}}}$$

□

Apart from the identities in law (8.31) and (8.32), we shall also use the much easier identity in law:

$$C|N| \stackrel{(\text{law})}{=} N|C| \ , \tag{8.33}$$

where C is a standard Cauchy variable, independent of N. We take up again the expression in (8.30), and we obtain

$$
\begin{aligned}
E\left[\exp\left(-\xi Z_{2r}\sqrt{Z_{\frac{1}{2},r}}\right)\right] &= E\left[\exp\left(-\xi|N|\sqrt{2Z_r}\right)\right] \ , && \text{by (8.32)} \\
&= E\left[\exp\left(i\xi C|N|\sqrt{2Z_r}\right)\right] \\
&= E\left[\exp\left(i\xi N|C|\sqrt{2Z_r}\right)\right] \ , && \text{by (8.33)} \\
&= E\left[\exp(-\xi^2 C^2 Z_r)\right] = E\left[\frac{1}{(1+\xi^2 C^2)^r}\right] \ .
\end{aligned}
$$

Thus, we obtain, with a constant c_μ which changes from line to line:

$$
\begin{aligned}
h(\xi) &= c_\mu \int_0^\infty \frac{du}{(1+u^2)(1+\xi^2 u^2)^r} = c_\mu \int_0^\infty \frac{dv}{\sqrt{v}(1+v)(1+v\xi^2)^r} \\
&= c_\mu \xi \int_0^1 \frac{dz \, z^{-1/2}(1-z)^{r-1/2}}{z+\xi^2(1-z)} \ ,
\end{aligned}
$$

with the change of variables: $v\xi^2 = \dfrac{z}{1-z}$.

Hence, going back to the definition of $h(\xi)$, we remark that we have obtained the identity:

$$E\left[\frac{1-A_1}{A_1+\xi^2(1-A_1)}\right] = E\left[\frac{1-Z}{Z+\xi^2(1-Z)}\right] \ ,$$

where $Z \stackrel{(\text{law})}{=} Z_{\frac{1}{2},\frac{1}{2\mu}}$, which proves the desired result:

$$A_1 \stackrel{(\text{law})}{=} Z \ .$$

(8.4.3)   We now prove the second result of F. Petit, i.e.:

$$A_1' \equiv A_{g_1} \stackrel{(\text{law})}{=} Z_{\frac{1}{2},\frac{1}{2}+\frac{1}{2\mu}}$$

Using the identities (8.25) and (8.27), we are able to compute the following quantity:

$$\gamma_{\theta,\lambda}' \stackrel{\text{def}}{=} \left[\exp\left(-\frac{\lambda^2}{2}A_{S_\theta}'\right)\right] \ .$$

We obtain:

$$\gamma'_{\theta,\lambda} = \frac{\theta}{\mu} \int_0^\infty db \, (\mathrm{ch}(\nu b) + \xi \mathrm{sh}(\nu b))^{-\frac{1}{\mu}} = \frac{\xi}{\mu} \int_0^\infty dx (\mathrm{ch}x + \xi \mathrm{sh}x)^{-\frac{1}{\mu}}$$

On the other hand, from the scaling property of $(A'_t, t \geq 0)$, we also obtain:

$$\gamma'_{\theta,\lambda} = E\left[\frac{\theta^2}{\theta^2 + \lambda^2 A'_1}\right] = E\left[\frac{\xi^2}{A'_1 + \xi^2(1 - A'_1)}\right] \quad .$$

Hence, we have obtained the following formula:

$$E\left[\frac{1}{A'_1 + \xi^2(1 - A'_1)}\right] = \frac{1}{\xi\mu} \int_0^\infty \frac{dx}{(\mathrm{ch}x + \xi\mathrm{sh}x)^{\frac{1}{\mu}}} \qquad (8.34)$$

In order to prove the desired result, we shall now use formula (8.29), which will enable us to make almost no computation.

In the case $\mu < 1$, we can define $\tilde{\mu} > 0$ by the formula: $\frac{1}{\mu} = 1 + \frac{1}{\tilde{\mu}}$, and we write $\tilde{A}_1$, for $A_1^{\tilde{\mu},-}$.

Hence, comparing formulae (8.29) and (8.34), we obtain:

$$E\left[\frac{1}{A'_1 + \xi^2(1 - A'_1)}\right] = \frac{\tilde{\mu}}{\mu} E\left[\frac{1 - \tilde{A}_1}{\tilde{A}_1 + \xi^2(1 - \tilde{A}_1)}\right] \qquad (8.35)$$

Now, since $\tilde{A}_1 \overset{(\mathrm{law})}{=} Z_{\frac{1}{2},\frac{1}{2\tilde{\mu}}}$, it is easily deduced from (8.35) that:

$$A'_1 \overset{(\mathrm{law})}{=} Z_{\frac{1}{2},\frac{1}{2\tilde{\mu}}+1} \quad ,$$

and since: $\frac{1}{2\tilde{\mu}} + 1 = \frac{1}{2} + \frac{1}{2\mu}$, we have shown, at least in the case $\mu < 1$:

$$A'_1 \overset{(\mathrm{law})}{=} Z_{\frac{1}{2},\frac{1}{2}+\frac{1}{2\mu}} \quad ,$$

which is F. Petit's second result.

(8.4.4) With a very small amount of extra computation, it is possible to extend P. Lévy's result even further, by considering, for given $\alpha, \beta > 0$:

$$A_t \equiv A_t^{\alpha,\beta} = \int_0^t ds \, 1_{(-\alpha\ell_s \leq B_s \leq \beta\ell_s)} \quad .$$

Indeed, taking up the above computation again, F. Petit has obtained the following extension of formula (8.29):

$$E\left[\frac{2\xi(1-A_1)}{A_1+\xi^2(1-A_1)}\right] = \int_0^\infty ds \frac{\varphi_\alpha(s)+\varphi_\beta(s)}{(\varphi_\alpha(s))^{1+\frac{1}{2\alpha}}(\varphi_\beta(s))^{1+\frac{1}{2\beta}}}$$

where we denote by $\varphi_a(s)$ the following quantity (which depends on ξ):

$$\varphi_a(s) \equiv \varphi_a^{(\xi)}(s) = \mathrm{ch}(as) + \xi\mathrm{sh}(as) \ .$$

8.5 A stochastic calculus approach to F. Petit's results

(8.5.1)   The main aim of this paragraph is to show, with the help of some arguments taken from stochastic calculus, the independence of the process $(A^{\mu,-}_{\tau^\mu(t)}, t \geq 0)$ and of the random variable $\ell^\mu_{\alpha^\mu_1,+}$, which, following the method discussed in the subparagraph (8.3.1), allows to reduce the computation of the law of $A^{\mu,-}_1$ to that of the pair $(A^{\mu,-}_{\tau^\mu_1}, A^{\mu,+}_{\tau^\mu_1})$, already presented in (8.14).

Since μ is fixed throughout the paragraph, we shall use the following simplified notation:

$$X_t = |B_t| - \mu\ell_t \ , \qquad X_t^+ = \sup(X_t, 0) \ ,$$

$$X_t^- = \sup(-X_t, 0) \ , \qquad A_t^\pm = \int_0^t ds \, 1_{(\pm X_s > 0)} \ ,$$

$(\ell^\mu_t, t \geq 0)$ denotes the local time at 0 of X, and $(\tau^\mu_t, t \geq 0)$ its right-continuous inverse.

(8.5.2)   We shall now adapt the stochastic calculus method developed by Pitman-Yor [75] to prove Lévy's arc sine law.

Tanaka's formula implies:

$$X_t^+ = M_t^{(+)} + \frac{1}{2}\ell^\mu_t \ , \qquad \text{where } M_t^{(+)} = \int_0^t 1_{(X_s > 0)}\mathrm{sgn}(B_s)dB_s \qquad (8.36)$$

$$X_t^- = -M_t^{(-)} - (1-\mu)\ell_t + \frac{1}{2}\ell_t^\mu \ , \text{ where } M_t^{(-)} = \int\limits_0^t 1_{(X_s<0)}\mathrm{sgn}(B_s)dB_s$$

Now, Knight's theorem about continuous orthogonal martingales allows to write:

$$M_t^{(+)} = \delta^{(+)}(A_t^+) \quad \text{and} \quad M_t^{(-)} = \delta^{(-)}(A_t^-) \ , \quad t \geq 0 \ ,$$

where $\delta^{(+)}$ and $\delta^{(-)}$ denote two independent Brownian motions, and the rest of the proof shall rely in an essential manner upon this independence result.

Using the time changes α^+ and α^-, the relations (8.36) become:

(i) $\qquad X_{\alpha_t^+}^+ = \delta_t^{(+)} + \frac{1}{2}\ell_{\alpha_t^+}^\mu;$

$$\text{(8.37)}$$

(ii) $\quad X_{\alpha_t^-}^- = -\delta_t^{(-)} - (1-\mu)\ell_{\alpha_t^-} + \frac{1}{2}\ell_{\alpha_t^-}^\mu$

The identity (i) in (8.37) may be interpreted as Skorokhod's reflection equation for the process $(X_{\alpha_t^+}^+, t \geq 0)$; hence, it follows that, just as in the case $\mu = 1$,

$$(X_{\alpha_t^+}^+, t \geq 0) \quad \textit{is a reflecting Brownian motion, and} \quad \frac{1}{2}\ell_{\alpha_t^+}^\mu = \sup_{s \leq t}(-\delta_s^{(+)})$$

$$\text{(8.38)}$$

In particular, we have: $\frac{1}{2}\ell_{\alpha_1^+}^\mu \overset{(\text{law})}{=} |N|$.

We now consider the identity (ii) in (8.37), which we write as:

$$X_{\alpha_t^-}^- = -Y_t^\mu + \frac{1}{2}\ell_{\alpha_t^-}^\mu \tag{8.39}$$

where:

$$Y_t^\mu \overset{\text{def}}{=} \delta_t^{(-)} + (1-\mu)\ell_{\alpha_t^-} \tag{8.40}$$

and we deduce from (8.39) that:

$$\frac{1}{2}\ell_{\alpha_t^-}^\mu = \sup_{s \leq t}(Y_s^\mu) \overset{\text{def}}{=} S_t^\mu \tag{8.41}$$

Hence, we have: $A_{\tau_{2t}^-}^- = \inf\left\{s : \frac{1}{2}\ell_{\alpha_s}^\mu > t\right\} = \inf\{s : Y_s^\mu > t\}$, from (8.39), and, in order to obtain the desired independence result, it suffices to prove that the process $(Y_t^\mu, t \geq 0)$ is measurable with respect to $(\delta_t^{(-)}, t \geq 0)$.

(8.5.3)   To prove this measurability result, we shall first express the process $(\ell_{\alpha_t^-}, t \geq 0)$ in terms of $\delta^{(-)}$ and Y^μ, which will enable us to transform the identity (8.40) into an equation, where Y^μ is the unknown process, and $\delta^{(-)}$ is the driving Brownian motion.

Indeed, if we consider again the identity (ii) in (8.37), we see that:

$$-\left(|B_{\alpha_t^-}| - \mu \ell_{\alpha_t^-}\right) = -X_{\alpha_t^-} = -\delta_t^{(-)} - (1 - \mu)\ell_{\alpha_t^-} + \frac{1}{2}\ell_{\alpha_t}^\mu \;,$$

which gives:

$$|B_{\alpha_t^-}| = \delta_t^{(-)} - \frac{1}{2}\ell_{\alpha_t^-}^\mu + \ell_{\alpha_t^-} \;, \qquad t \geq 0 \;.$$

Again, this equality may be considered as an example of Skorokhod's reflection equation for the process $(|B_{\alpha_t^-}|, t \geq 0)$. Therefrom, we deduce:

$$\ell_{\alpha_t^-} = \sup_{s \leq t}\left(-\delta_s^{(-)} + \frac{1}{2}\ell_{\alpha_s}^\mu\right) = \sup_{s \leq t}\left(-\delta_s^{(-)} + S_s^\mu\right) \;, \qquad \text{using (8.41).}$$

Bringing the latter expression of $\ell_{\alpha_t^-}$ into (8.40), we obtain:

$$Y_t^\mu = \delta_t^{(-)} + (1 - \mu)\sup_{s \leq t}(-\delta_s^{(-)} + S_s^\mu) \tag{8.42}$$

Now, in the case $\mu \in \,]0, 2[$, the fixed point theorem allows to show that this equation admits one and only one solution $(Y_t^\mu, t \geq 0)$, and that this solution is adapted with respect to $(\delta_t^{(-)}, t \geq 0)$.

Indeed, the application:

$$\Phi : \Omega_{0,T}^* \equiv \{f \in C([0, T]; \mathbb{R}); f(0) = 0\} \longrightarrow \Omega_{0,T}^*$$

$$g \longrightarrow \left(\delta_t^{(-)} + (1 - \mu)\sup_{s \leq t}\left(-\delta_s^{(-)} + \sup_{u \leq s}(g(u))\right); t \leq T\right)$$

is Lipschitz, with coefficient $K = |1 - \mu|$, i.e.:

$$\sup_{t \leq T}|\Phi(g)(t) - \Phi(h)(t)| \leq K \sup_{t \leq T}|g(t) - h(t)| \;.$$

Hence, if $\mu \in]0, 2[$, Φ is strictly contracting, and Picard's iteration procedure converges, therefore proving at the same time the uniqueness of the solution of (8.42) and its measurability with respect to $\delta^{(-)}$.

Remark: The difficulty to solve (8.42) when μ does not belong to the interval $]0, 2[$ was already noticed in Le Gall-Yor [62], and partly dealt with there.

Comments on Chapter 8

A number of extensions of Lévy's arc sine law for Brownian motion have been presented in this chapter, with particular emphasis on F. Petit's results (8.5) and (8.6). The paragraph 8.4, and particularly the subparagraph (8.4.2), is an attempt to explain the results (8.5) and (8.6), using the extension of the Ray-Knight theorems proved in Chapter 3 for the process $(|B_t| - \mu \ell_t; t \leq \tau_s)$. In the next Chapter, another explanation of (8.5) is presented.

Chapter 9

Further results about reflecting Brownian motion perturbed by its local time at 0

In this Chapter, we study more properties of the process

$$(X_t \equiv |B_t| - \mu \ell_t, t \geq 0)$$

which played a central role in the preceding Chapter 8. One of the main aims of the present Chapter is to give a clear proof of the identity in law between the pairs (8.14) and (8.16), that is:

$$\frac{1}{t^2} \left(A^{\mu,-}_{\tau^\mu_t}, A^{\mu,+}_{\tau^\mu_t} \right) \stackrel{\text{(law)}}{=} \frac{1}{8} \left(\frac{1}{Z_{\frac{1}{2\mu}}}, \frac{1}{Z_{\frac{1}{2}}} \right) \tag{9.1}$$

(recall that $(\tau^\mu_t, t \geq 0)$ is the inverse of the local time $(\ell^\mu_t, t \geq 0)$ at 0 for the process X.)

9.1 A Ray-Knight theorem for the local times of X, up to τ^μ_s, and some consequences

The main result of this Chapter is the following

Theorem 9.1 *Fix $s > 0$. The processes $(\ell^x_{\tau^\mu_s}(X); x \geq 0)$ and $(\ell^{-x}_{\tau^\mu_s}(X); x \geq 0)$ are independent, and their respective laws are Q^0_s and $Q^{2-\frac{2}{\mu}}_s$, where $Q^{2-\frac{2}{\mu}}_s$ denotes the law of the square, starting from s, of the Bessel process with dimension $2 - \frac{2}{\mu}$, and absorbed at 0.*

Corollary 9.1.1 *We have the following identities in law:*

a) $\mu\ell_{\tau_s^\mu} \equiv -\inf\{X_u; u \le \tau_s^\mu\} \overset{\text{(law)}}{=} \dfrac{s}{2Z_{\frac{1}{\mu}}}$; b) $\sup\{X_u; u \le \tau_s^\mu\} \overset{\text{(law)}}{=} \dfrac{s}{2Z_1}$

c) $A_{\tau_s^\mu}^{\mu,-} \overset{\text{(law)}}{=} \dfrac{s^2}{8Z_{\frac{1}{2\mu}}}$; d) $A_{\tau_s^\mu}^{\mu,+} \overset{\text{(law)}}{=} \dfrac{s^2}{8Z_{\frac{1}{2}}}$.

Moreover, the pairs $\left(\mu\ell_{\tau_s^\mu}, A_{\tau_s^\mu}^{\mu,-}\right)$ and $\left(\sup\{X_u; u \le \tau_s^\mu\}, A_{\tau_s^\mu}^{\mu,+}\right)$ are independent.

In particular, the identity in law (9.1) holds.

PROOF OF THE COROLLARY:

1) The independence statement follows immediately from the independence of the local times indexed by $x \in \mathbb{R}_+$, and $x \in \mathbb{R}_-$, as stated in Theorem 9.1.

2) We prove a). Remark that:

$$-\mu\ell_{\tau_s^\mu} = \inf\{X_u; u \le \tau_s^\mu\} = \inf\left\{x \in \mathbb{R}; \ell_{\tau_s^\mu}^x(X) > 0\right\} \ ;$$

hence, from Theorem 9.1, we know that the law of $\mu\ell_{\tau_s^\mu}$ is that of the first hitting time of 0, by a $\text{BESQ}_s^{2-\frac{2}{\mu}}$ process, which implies the result a), using time reversal. The same arguments, used with respect to the local times $\left(\ell_{\tau_s^\mu}^x(X); x \ge 0\right)$, give a proof of b).

3) In order to prove c), we first remark that, by scaling, we can take $s = 1$. Then, we have:

$$A_{\tau_1^\mu}^{\mu,-} = \int_0^\infty dy\, \ell_{\tau_1^\mu}^{-y}(X) \overset{\text{(law)}}{=} \int_0^{\mathcal{L}_1} dy\, Y_y \ ,$$

where $(Y_y; y \ge 0)$ is a $\text{BESQ}_0^{2+\frac{2}{\mu}}$ process, using Theorem 9.1, and time-reversal, and $\mathcal{L}_1 = \sup\{y : Y_y = 1\}$.

We now use the following result on powers of BES-processes (see Biane-Yor [17]):

$$qR_\nu^{1/q}(t) = R_{\nu q}\left(\int_0^t ds\, R_\nu^{-2/p}(s)\right) \ , \qquad t \ge 0 \qquad (9.2)$$

where (R_λ) is a BES process with index λ, and $\frac{1}{p}+\frac{1}{q}=1$. We take $p=-1$, and $q=\frac{1}{2}$. We then deduce from (9.2) that:

$$\int\limits_0^{\mathcal{L}_1(R_\nu)} ds\, R_\nu^2(s) = \mathcal{L}_{1/2}(R_{\nu/2}) \stackrel{\text{(law)}}{=} \frac{1}{4}\mathcal{L}_1(R_{\nu/2}) \stackrel{\text{(law)}}{=} \frac{1}{8}\frac{1}{Z_{\nu/2}}$$

and c) follows by taking $\nu=\dfrac{1}{\mu}$.

d) follows similarly, by considering $\left(\ell^x_{\tau_s^\mu}(X); x\geq 0\right)$ and $\nu=1$. $\square$

Using again Theorem 9.1 and the identity (9.2) in conjunction, we obtain, at no extra cost, the following extension of Corollary 9.1.1.

Corollary 9.1.2 *Let* $\alpha\geq 0$. *We have:*

$$\left\{\int\limits_{-\infty}^0 dy\left(\ell^y_{\tau_s^\mu}(X)\right)^\alpha ;\, \int\limits_0^\infty dy\left(\ell^y_{\tau_s^\mu}(X)\right)^\alpha\right\} \stackrel{\text{(law)}}{=} \frac{s^{\alpha+1}}{2(1+\alpha)^2}\left(\frac{1}{Z_{\frac{1}{\mu(1+\alpha)}}};\, \frac{1}{Z_{\frac{1}{1+\alpha}}}\right)$$

$$\tag{9.3}$$

where, on the right-hand side, the two gamma variables are independent.

Remark: In order to understand better the meaning of the quantities on the left-hand side of (9.3), it may be interesting to write down the following equalities, which are immediate consequences of the occupation density formula for X;
let $\varphi:\mathbb{R}\to\mathbb{R}_+$, and $h:\mathbb{R}_+\to\mathbb{R}_+$, be two Borel functions; then, the following equalities hold:

$$\int\limits_0^t du\,\varphi(X_u)h(\ell^{X_u}_t) = \int\limits_{-\infty}^\infty dy\,\varphi(y)h(\ell^y_t)\ell^y_t$$

$$\int\limits_0^t du\,\varphi(X_u)h(\ell^{X_u}_u) = \int\limits_{-\infty}^\infty dy\,\varphi(y)H(\ell^y_t)\ ,\qquad \text{where: } H(x)=\int\limits_0^x dz\, h(z)\ .$$

In particular, if we take: $h(x)=x^{\alpha-1}$, for $\alpha>0$, we obtain:

$$\int\limits_{-\infty}^\infty dy\,\varphi(y)(\ell^y_t)^\alpha = \int\limits_0^t du\,\varphi(X_u)(\ell^{X_u}_t)^{\alpha-1} = \alpha\int\limits_0^t du\,\varphi(X_u)(\ell^{X_u}_u)^{\alpha-1}\ .$$

Exercise 9.1 Prove the following extension of the Földes-Révész identity in law (4.11):

for $s \geq q$,

$$\int_0^\infty dy\, 1_{(0 < \ell_{\tau_s}^{-y}(X) < q)} \overset{(\text{law})}{=} T_{\sqrt{q}}\left(R_{\frac{2}{\mu}}\right) . \tag{9.4}$$

9.2 Proof of the Ray-Knight theorem for the local times of X

(9.2.1) In order to prove Theorem 9.1, it is important to be able to compute expressions such as:

$E\left[\exp(-H_{\tau_s^\mu})\right]$, where: $H_t = \int_0^t ds\, h(X_s)$, with $h : \mathbb{R} \to \mathbb{R}_+$ a Borel func-

tion. The fact that $(H_t, t \geq 0)$ is an additive functional of the Markov process $\{Z_t = (B_t, \ell_t); t \geq 0\}$ shall play an important role in the sequel.

To have access to the above quantity, we shall consider in fact:

$$\gamma = E\left[\int_0^\infty ds \exp\left(-\frac{\theta^2 s}{2}\right) \exp(-H_{\tau_s^\mu})\right]$$

and then, after some transformations, we shall invert the Laplace transform in $\frac{\theta^2}{2}$.

(9.2.2) From now on, we shall use freely the notation and some of the results in Biane-Yor [17] and Biane [14], concerning Brownian path decomposition; in particular, we shall use *Bismut's identity:*

$$\int_0^\infty dt\, P_0^t = \int_0^\infty ds\, P^{\tau(s)} \circ \int_{-\infty}^\infty da\, {}^\vee(P_a^{T_0})$$

which may be translated as:

$$\int_0^\infty dt\, E\left[F(B_u, u \le g_t)G(B_{t-v}; v \le t - g_t)\right]$$

$$= \int_0^\infty ds\, E\left[F(B_u; u \le \tau_s)\right] \int_{-\infty}^\infty da\, E_a\left[G(B_h, h \le T_0)\right]$$

(9.5)

where F and G are two measurable, $\mathbb{R}_+$-valued, Brownian functionals. Here is an important application of formula (9.5):

if we consider $C_t = \int_0^t du\varphi(B_u, \ell_u)$, where φ is an $\mathbb{R}_+$-valued continuous function, and $f : \mathbb{R} \times \mathbb{R}_+ \to \mathbb{R}_+$ is another continuous function, then:

$$E\left[\int_0^\infty du\, f(B_u, \ell_u)\exp(-C_u)\right]$$

$$= \int_0^\infty ds \int_{-\infty}^\infty da\, f(a, s)E_0\left[\exp(-C_{\tau_s})\right] E_a\left[\exp -C_{T_0}^s\right]$$

(9.6)

where $C_t^s = \int_0^t du\varphi(B_u, s)$.

(9.2.3) We are now ready to transform γ. First, we write:

$$\gamma = E\left[\int_0^\infty d\ell_u^\mu \exp\left(-\frac{\theta^2}{2}\ell_u^\mu\right)\exp(-H_u)\right]$$

$$= \lim_{\varepsilon \to 0} \frac{1}{\varepsilon} \int_{-\infty}^\infty da \int_0^\infty ds\, 1_{(0 \le |a| - \mu s \le \varepsilon)}g(s)k(a, s)$$

(9.7)

where:

$$g(s) = E\left[\exp\left(-\frac{\theta^2}{2}\ell_{\tau_s}^\mu\right)\exp(-H_{\tau_s})\right]$$

$$k(a, s) = E_a\left[\exp\left(-\frac{\theta^2}{2}\ell_{T_0}^\mu - \int_0^{T_0} du\, h(|B_u| - \mu s)\right)\right] ,$$

with $\ell^\mu_{T_0}$ denoting the local time at 0 of $(|B_u| - \mu s; u \le T_0)$.

From (9.7), it easily follows that:

$$\gamma = \int_{-\infty}^{\infty} db \; g(|b|)k(b\mu, |b|) = 2 \int_0^{\infty} db \; g(b)k(b\mu, b) \ .$$

It is now natural to introduce $\varphi_b(x)dx$, the law of $\ell^\mu_{T_b}$, resp.: $\psi_a(y)dy$, the law of $\ell^a_{T_0}$ under P_a, as well as the conditional expectations:

$$e^{(1)}(b, x) = E\left[\exp(-H_{T_b}) \mid \ell^\mu_{T_b} = x\right]$$

$$e^{(2)}(a, y) = E_a\left[\exp\left(-\int_0^{T_0} du \; h(|B_u| - a)\right) \mid \ell^a_{T_0} = y\right] \ .$$

These notations enable us to write γ as follows:

$$\gamma = 2 \int_0^{\infty} db \int_0^{\infty} dx \varphi_b(x) \exp\left(-\frac{\theta^2 x}{2}\right) e^{(1)}(b, x) \int_0^{\infty} dy \psi_{b\mu}(y) \exp\left(-\frac{\theta^2 y}{2}\right) e^{(2)}(b\mu, y) \ .$$

It is now easy to invert the Laplace transform, and we get:

$$E\left[\exp(-H_{\tau^\mu_s})\right] = 2 \int_0^{\infty} db \int_0^{s} dx \varphi_b(x) e^{(1)}(b, x) \psi_{b\mu}(s-x) e^{(2)}(b\mu, s-x) \ . \quad (9.8)$$

Plainly, one would like to be able to disintegrate the above integral with respect to $db\,dx$, and, tracing our steps back, we arrive easily, with the help of Bismut's decomposition to the following reinforcement of (9.8):

$$E\left[\exp(-H_{\tau^\mu_s}) \mid \ell_{\tau^\mu_s} = b, \ell^\mu_{g_{\tau^\mu_s}} = x\right] = e^{(1)}(b, x)e^{(2)}(b\mu, s - x) \ ,$$

and:

$$P\left\{\ell_{\tau^\mu_s} \in db, \ell^\mu_{g_{\tau^\mu_s}} \in dx\right\} = 2db \; dx \varphi_b(x)\psi_{\mu b}(s - x)1_{(x \le s)} \ .$$

However, we know, from Chapter 3, the explicit expressions of $\varphi_b(x)$ and $\psi_a(y)$; this implies the following

Proposition 9.1 *For fixed s, the variables $\ell_{\tau^\mu_s}$ and $\ell^\mu_{g_{\tau^\mu_s}}$ are independent and they satisfy:*

$$\mu \ell_{\tau^\mu_s} \overset{\text{(law)}}{=} \frac{s}{2Z_{\frac{1}{\mu}}} \overset{\text{(law)}}{=} \frac{s}{\ell^\mu_{\tau_1}} \ ; \quad \ell^\mu_{g_{\tau^\mu_s}} \overset{\text{(law)}}{=} s \; Z_{\frac{1}{\mu},1}$$

(9.2.4)   We are now in a position to prove Theorem 9.1, as we know how to write $e^{(1)}(b,x)$ and $e^{(2)}(b\mu, s-x)$ in terms of the laws of BESQ processes of different dimensions.

We first recall that, in Chapter 3, we proved the following RK theorem (Theorem 3.4):
$\{\ell_{\tau_b}^{a-\mu b}(X); a \geq 0\}$ is an inhomogeneous Markov process, which is $BESQ_0^{2/\mu}$, for $a \leq \mu b$, and $BESQ^0$, for $a \geq \mu b$.

Hence, we may write:

$$
e^{(1)}(b,x) = Q_x^0 \left(\exp - \int_0^\infty dz\, h(z) Y_z \right) \dots
$$

$$
\dots Q_0^{2/\mu} \left(\exp - \int_0^{\mu b} dz\, h(z - \mu b) Y_z \mid Y_{\mu b} = x \right)
$$

$$
e^{(2)}(b\mu, s-x) = Q_{s-x}^0 \left(\exp - \int_0^\infty dz\, h(z) Y_z \right) \dots
$$

$$
\dots Q_0^2 \left(\exp - \int_0^{\mu b} dz\, h(z - \mu b) Y_z \mid Y_{\mu b} = s - x \right)
$$

Therefore, the product of these two expressions is equal, thanks to the additivity properties of $\{Q_s^0\}$ and $\{Q_0^\delta\}$, to:

$$
e(b,x,s)
$$

$$
= Q_s^0 \left(\exp - \int_0^\infty dz\, h(z) Y_z \right) Q_0^{2+\frac{2}{\mu}} \left(\exp - \int_0^{\mu b} dz\, h(z - \mu b) Y_z \mid Y_{\mu b} = s \right)
$$

and we make the important remark that this expression no longer depends on x.

Putting together the different results we have obtained up to now, we can state the following

Theorem 9.2 1) The process $\left\{ \ell_{\tau_s^\mu}^x(X); x \in \mathbb{R} \right\}$ is independent of the variable $\ell_{g_{\tau_s^\mu}}^\mu$;

2) The processes $\left\{\ell^x_{\tau^\mu_s}(X); x \geq 0\right\}$ and $\left\{\ell^{-x}_{\tau^\mu_s}(X); x \geq 0\right\}$ are independent;

3) The law of $\left\{\ell^x_{\tau^\mu_s}(X); x \geq 0\right\}$ is Q^0_s;

4) The law of $\left\{\ell^{y-\mu b}_{\tau^\mu_s}(X); 0 \leq y \leq \mu b\right\}$ is $Q^{2+\frac{2}{\mu}}_{0 \xrightarrow[(\mu b)]{} s}$.

(9.2.5)   We now end the proof of Theorem 9.1, by remarking that, from Proposition 9.1, $T_0 \equiv \inf\left\{x : \ell^{-x}_{\tau^\mu_s}(X) = 0\right\} = \mu \ell_{\tau^\mu_s}$ is distributed as T_0 under $Q^{2-\frac{2}{\mu}}_s$, and that when we reverse the process:
$(\ell^{-y}_{\tau^\mu_s}; 0 \leq y \leq T_0)$ from $T_0 \equiv \mu b$, that is, we consider:
$\left\{\ell^{-(\mu b - x)}_{\tau^\mu_s} \equiv \ell^{x-\mu b}_{\tau^\mu_s}; 0 \leq x \leq \mu b\right\}$ conditioned on $T_0 = \mu b$, we find that the latter process is distributed as $Q^{2+\frac{2}{\mu}}_{0 \xrightarrow[(\mu b)]{} s}$.

Putting together these two results, we find that

$$\left\{\ell^{-x}_{\tau^\mu_s}(X); x \geq 0\right\} \text{ is distributed as } \mathrm{BESQ}^{2-\frac{2}{\mu}}_s ,$$

since it is well-known that:

$$\left(R^{(\nu)}_0(L_s - u); u \leq \mathcal{L}_s\right) \overset{\text{(law)}}{=} \left(R^{(-\nu)}_s(u); u \leq T_0\right)$$

where $(R^{(\alpha)}_a(t); t \geq 0)$ denotes here the Bessel process with index α, starting at a,

$$\mathcal{L}_s = \sup\left\{t; R^{(\nu)}_0(t) = s\right\} , \quad \text{and} \quad T_0 = \inf\left\{t; R^{(-\nu)}_s(t) = 0\right\}$$

9.3 Generalisation of a computation of F. Knight

(9.3.1)   In his article in the Colloque Paul Lévy (1987), F. Knight [58] proved the following formula:

$$E\left[\exp\left(-\frac{\lambda^2}{2} \frac{A^+_{\tau_s}}{M^2_{\tau_s}}\right)\right] = \frac{2\lambda}{\mathrm{sh}(2\lambda)} , \quad \lambda \in \mathbb{R} , \; s > 0 , \tag{9.9}$$

where $A_t^+ = \int\limits_0^t ds\, 1_{(B_s>0)}$, $M_t = \sup\limits_{u\leq t} B_u$, and $(\tau_s, s \geq 0)$ is the inverse of the local time of B at 0.

Time changing $(B_t^+, t \geq 0)$ into a reflecting Brownian motion with the help of the well-known representation, already used in paragraph 4.1:

$$B_t^+ = \beta \left(\int_0^t ds\, 1_{(B_s>0)} \right) \ ,$$

where $(\beta(u), u \geq 0)$ denotes a reflecting Brownian motion, formula (9.9) may also be written in the equivalent form:

$$E\left[\exp\left(-\frac{\lambda^2}{2} \frac{\tau_s}{(M_{\tau_s}^*)^2} \right) \right] = \frac{2\lambda}{\mathrm{sh}(2\lambda)} \ , \tag{9.10}$$

where: $M_t^* = \sup\limits_{u\leq t} |B_u|$.

Formulae (9.9) and (9.10) show that:

$$\frac{A_{\tau_s}^+}{M_{\tau_s}^2} \overset{\text{(law)}}{=} \frac{\tau_s}{(M_{\tau_s}^*)^2} \overset{\text{(law)}}{=} T_2^{(3)} \overset{\text{def}}{=} \inf\{t : R_t = 2\} \ , \tag{9.11}$$

where $(R_t, t \geq 0)$ is a 3-dimensional Bessel process starting from 0.

An explanation of the identity in law (9.11) has been given by Ph. Biane [13] and P. Vallois [88], with the help of a pathwise decomposition.

For the moment, we generalize formulae (9.9) and (9.10) to the μ-process X, considered up to τ_s^μ.

Theorem 9.3 (We use the notation in the above paragraphs)

1) Define $I_u^\mu = \inf\limits_{v\leq u} X_v$. Then, we have:

$$E\left[\exp\left(-\frac{\lambda^2}{2} \frac{A_{\tau_s^\mu}^{\mu,-}}{(I_{\tau_s^\mu}^\mu)^2} \right) \right] = \left(\frac{\lambda}{\mathrm{sh}\lambda} \right) \left(\frac{1}{\mathrm{ch}\lambda} \right)^{1/\mu} . \tag{9.12}$$

2) Define: $X_t^* = \sup\limits_{s\leq t} |X_s|$. Then, if we denote $c = 1/\mu$, we have:

$$E\left[\exp\left(-\frac{\lambda^2}{2}\frac{\tau_s^\mu}{(X_{\tau_s^\mu}^*)^2}\right)\right] = \frac{1}{2^c}\left(\frac{\lambda}{\mathrm{sh}\,\lambda}\right)\left(\frac{1}{\mathrm{ch}\,\lambda}\right) + c\lambda(\mathrm{sh}\,\lambda)^{c-1}\int_\lambda^{2\lambda}\frac{du}{(\mathrm{sh}\,u)^{c+1}}$$

$$(9.13)$$

PROOF:

1) Recall that: $I_{\tau_s^\mu}^\mu = -\mu\ell_{\tau_s^\mu}$.

In order to prove formula (9.12), we first deduce from Theorem (9.2) that:

$$E\left[\exp\left(-\frac{\lambda^2}{2(\mu b)^2}\int_0^{\mu b} dy\,\ell_{\tau_s^\mu}^{y-\mu b}(X)\right)\,\Big|\,\ell_{\tau_s^\mu} = b\right]$$

$$= Q_0^{2+\frac{2}{\mu}}\left(\exp\left(-\frac{\lambda^2}{2(\mu b)^2}\int_0^{\mu b} dy\,Y_y\right)\,\Big|\,Y_{\mu b} = s\right)$$

Using Lévy's generalized formula (2.5), this quantity is equal to:

$$(*) \qquad\qquad \left(\frac{\lambda}{\mathrm{sh}\,\lambda}\right)^{1+\frac{1}{\mu}}\exp\left(-\frac{s}{2\mu b}(\lambda\coth\lambda - 1)\right) \quad .$$

Then, integrating with respect to the law of $\ell_{\tau_s^\mu}$ in the variable b, we obtain that the left-hand side of (9.12) is equal to:

$$\left(\frac{\lambda}{\mathrm{sh}\,\lambda}\right)^{1+\frac{1}{\mu}}\left(\frac{\mathrm{th}\,\lambda}{\lambda}\right)^{\frac{1}{\mu}} = \left(\frac{\lambda}{\mathrm{sh}\,\lambda}\right)\left(\frac{1}{\mathrm{ch}\,\lambda}\right)^{\frac{1}{\mu}} \quad . \qquad (9.14)$$

2) Formula (9.13) has been obtained by F. Petit and Ph. Carmona using the independence of $\left(\ell_{\tau_s^\mu}^x(X); x \geq 0\right)$ and $\left(\ell_{\tau_s^\mu}^{-x}(X); x \geq 0\right)$, as asserted in Theorem 9.1, together with the same kind of arguments as used in the proof of formula (9.12). □

The following exercise may help to understand better the law of $\dfrac{\tau_s^\mu}{(X_{\tau_s^\mu}^*)^2}$

Exercise 9.2 Let $c = \frac{1}{\mu} > 0$. Consider a pair of random variables (T, H) which is distributed as follows:

(i) H takes its values in the interval $[1,2]$, with:

$$P(H = 1) = \frac{1}{2^c} \; ; \quad P(H \in dx) = c\frac{dx}{x^{c+1}} \qquad (1 < x < 2)$$

(ii) For $\lambda > 0$, we have:

$$E\left[\exp\left(-\frac{\lambda^2}{2}T\right) \mid H = 1\right] = \left(\frac{\lambda}{\text{sh } \lambda}\right)\left(\frac{1}{\text{ch } \lambda}\right)^c \, ,$$

and, for $1 < x < 2$:

$$E\left[\exp\left(-\frac{\lambda^2}{2}T\right) \mid H = x\right] \underset{(a)}{=} \left(\frac{\lambda x}{\text{sh } \lambda x}\right)^2 \left(\frac{x \,\text{sh } \lambda}{\text{sh } \lambda x}\right)^{\frac{1}{\mu}-1}$$

$$\underset{(b)}{=} \left(\frac{\lambda x}{\text{sh } \lambda x}\right)^{\frac{1}{\mu}+1} \left(\frac{\lambda}{\text{sh } \lambda}\right)^{1-\frac{1}{\mu}}$$

(we present both formulae (a) and (b), since, in the case $\mu \leq 1$, the right-hand side of (a) clearly appears as a Laplace transform in $\frac{\lambda^2}{2}$, whereas in the case $\mu \geq 1$, the right-hand side of (b) clearly appears as a Laplace transform in $\frac{\lambda^2}{2}$).

Now, prove that:

$$\frac{T_s^{\mu}}{(X_{T_s^{\mu}}^*)^2} \overset{\text{(law)}}{=} T \tag{9.15}$$

We now look for some probabilistic explanation of the simplification which occurred in (9.14), or, put another way, what does the quantity $\left(\frac{\text{th}\lambda}{\lambda}\right)^{\frac{1}{\mu}}$ represent in the above computation?

With this in mind, let us recall that:

$$\mu \ell_{T_s^{\mu}} \overset{\text{(law)}}{=} \frac{s}{\ell_{T_1}^{\mu}} \, , \quad \text{and} \quad P\left(\ell_{T_1}^{\mu} \in dy\right) = Q_0^{2/\mu}(Y_1 \in dy) \, .$$

Thus, the integral with respect to (db) of the term featuring $(\lambda \coth \lambda - 1)$ in $(*)$, above (9.14), gives us:

$$\underline{R}^{\delta}\left(\exp -\frac{\lambda^2}{2}\int_0^1 dy \, Y_y\right) \, ,$$

where $\delta = \frac{2}{\mu}$, and we have used the notation in Chapter 3, concerning the decomposition:

$$Q_0^\delta = Q_{0\to 0}^\delta * \underline{R}^\delta .$$

Developing the same arguments more thoroughly, we obtain a new Ray-Knight theorem which generalizes formula (9.12).

Theorem 9.4 *For simplicity, we write $I = I_{\tau_s^\mu}^\mu$. We denote by $(\lambda_t^x(X); x \geq 0)$ the process of local times defined by means of the formula:*

$$\frac{1}{I^2} \int_0^t du\, f\left(1 - \frac{X_u}{I}\right) = \int_0^\infty dx\, f(x)\lambda_t^x(X) , \quad t \leq \tau_s^\mu ,$$

for every Borel function $f : [0,1] \to \mathbb{R}_+$.

Then, the law of $(\lambda_{\tau_s^\mu}^x; 0 \leq x \leq 1)$ is $Q_{0\to 0}^2 * Q_0^{2/\mu}$.

PROOF: By scaling, we can take $s = 1$. Using again Theorem 9.2, it suffices, in order to compute:

$$E\left[\exp - \frac{1}{I^2} \int_0^{\tau_1^\mu} ds\, f\left(1 - \frac{X_u}{I}\right)\right] ,$$

to integrate with respect to the law of $\ell_{\tau_1^\mu}$ the quantity:

$$Q_0^{2+\frac{2}{\mu}}\left(\exp - \frac{1}{(\mu b)^2} \int_0^{\mu b} dy\, f\left(\frac{y}{\mu b}\right) Y_y \mid Y_{\mu b} = 1\right)$$

$$= Q_0^{2+\frac{2}{\mu}}\left(\exp - \int_0^1 dz\, f(z) Y_z \mid Y_1 = \frac{1}{\mu b}\right)$$

$$= (\ell(f))^{1+\frac{1}{\mu}} \exp\left(-\frac{1}{2\mu b} h(f)\right) ,$$

where $\ell(f)$ and $h(f)$ are two constants depending only on f.

When we integrate with respect to the law of $\mu \ell_{\tau_1^\mu}$, which is that of $\frac{1}{Y_1}$ under $Q_0^{2/\mu}$, we find:

$$\ell(f)\left(\frac{\ell(f)}{1 + h(f)}\right)^{\frac{1}{\mu}} ,$$

which is equal to the expectation of $\exp\left(-\int_0^1 dy\, f(y)Y_y\right)$ under

$Q_{0 \to 0}^2 * Q_0^{2/\mu}.$ □

Remark: A more direct proof of Theorem 9.4 may be obtained by using jointly Theorem 9.1 together with Corollary 3.9.2, which expresses the law of a Bessel process transformed by Brownian scaling relative to a last passage time.

9.4 Towards a pathwise decomposition of $(X_u; u \leq \tau_s^\mu)$

In order to obtain a more complete picture of $(X_u; u \leq \tau_s^\mu)$, we consider again the arguments developed in paragraph 9.2, but we now work at the level of Bismut's identity (9.5) itself, instead of remaining at the level of the local times of X, as was done from (9.7) onwards.

Hence, if we now define, with the notation introduced in (9.5):

$$\gamma = E\left[\int_0^\infty ds \, \exp\left(-\frac{\theta^2 s}{2}\right) \Phi_{\tau_s^\mu}\right] ,$$

where

$$\Phi_t = F(B_u, u \leq g_t)G(B_{t-v} : v \leq t - g_t) ,$$

we obtain, with the same arguments as in paragraph 9.2:

$$E\left[F(B_u : u \leq g_{\tau_s^\mu})G(B_{\tau_s^\mu - v}; v \leq \tau_s^\mu - g_{\tau_s^\mu}) \mid \ell_{\tau_s^\mu} = b, \ell_{g_{\tau_s^\mu}}^\mu = x\right]$$
$$= E\left[F(B_u; u \leq \tau_b) \mid \ell_{\tau_b}^\mu = x\right] E_{b\mu}\left[G(B_h; h \leq T_0) \mid \ell_{T_0}^{b\mu} = s - x\right] \quad (9.16)$$

which may be translated in the form of the integral representation:

$$P^{\tau_s^\mu} = 2\int_0^\infty db \int_0^s dx \varphi_b(x)\psi_{b\mu}(s-x)P^{\tau_b}\left(\cdot \mid \ell_{\tau_b}^\mu = x\right) \circ^\vee \left(P_{b\mu}^{T_0}\right)\left(\cdot \mid \ell_{T_0}^{b\mu} = s - x\right)$$

or, even more compactly:

$$\int\limits_{0}^{\infty} ds \, P^{\tau_s^{\mu}} = 2 \int\limits_{0}^{\infty} db \, P^{T_b} \circ {}^{\vee} \left(P_{b\mu}^{T_0} \right) . \tag{9.17}$$

For the moment, we simply deduce from (9.16) that, conditionally on $\ell_{\tau_s^{\mu}} = b$, the process:

$$(X_v; v \leq \tau_s^{\mu} - g_{\tau_s^{\mu}}) \equiv (B_{\tau_s^{\mu} - v} - \mu \ell_{\tau_s^{\mu}}; v \leq \tau_s^{\mu} - g_{\tau_s^{\mu}})$$
$$= (B_{\tau_s^{\mu} - v} - \mu b; v \leq \tau_s^{\mu} - g_{\tau_s^{\mu}})$$

is distributed as Brownian motion starting from 0, considered up to its first hitting time of $-\mu b$.

It would now remain to study the pre-$g_{\tau_s^{\mu}}$ process in the manner of Biane [13] and Vallois [88], but this is left to the diligent reader.

Comments on Chapter 9

The results presented in this Chapter were obtained by the second author while teaching the course at ETH (Sept. 91–Feb. 92). In the end, Theorem 9.1 may be used to give, thanks to the scaling property, a quick proof of F. Petit's first result (8.5).

The main difference between Chapter 8 and the present Chapter is that, in Chapter 8, the proof of F. Petit's first result (8.5) was derived from a Ray-Knight theorem for the local times of X, considered up to $\tau_s = \inf\{t : \ell_t = s\}$, whereas, in the present chapter, this result (8.5) is obtained as a consequence of Theorem 9.1, which is a RK theorem for the local times of X, up to $\tau_s^{\mu} \equiv \inf\{t : \ell_t^{\mu} = s\}$, a more intrinsic time for the study of X.

As a temporary conclusion on this topic, it may be worth to emphasize the simplicity (in the end!) of the proof of (8.5):

- it was shown in (8.3.1) that a proof of the arc sine law for Brownian motion may be given in a few moves, which use essentially two ingredients:

(i) the scaling property,

(ii) the independence, and the identification of the laws, of $A_{\tau(1)}^+$ and $A_{\tau(1)}^-$, the latter being, possibly, deduced from excursion theory;

- to prove F. Petit's result, the use of the scaling property makes no problem, whilst the independence and the identification of the laws of $A_{\tau^\mu(1)}^{\mu,-}$ and $A_{\tau^\mu(1)}^{\mu,+}$ are dealt with in Theorem 9.1.

However, the analogy with the Brownian case is not quite complete, since we have not understood, most likely from excursion theory, the identity in law between the quantities (8.13) and (8.14), as done in Pitman-Yor [77] and Perman-Pitman-Yor [69] in the Brownian case.

Chapter 10

On principal values of Brownian and Bessel local times

In real and complex analysis, the Hilbert transform H, which may be defined, for any $f \in L^2(\mathbb{R})$, as:

$$Hf(x) = \frac{1}{\pi} \lim_{\varepsilon \to 0} \int_{-\infty}^{\infty} \frac{dy\, f(y)}{y - x} 1_{(|y-x| \geq \varepsilon)} \qquad (10.1)$$

(this limit exists dx a.s.)

plays an important role, partly because of the fundamental identity between Fourier transforms:

$$\widehat{Hf}(\xi) = i\, \mathrm{sgn}(\xi)\hat{f}(\xi)$$

If, in (10.1), f is assumed to be Hölder continuous, and has compact support, then the limit in ε exists for every $x \in \mathbb{R}$. This remark applies to $f(y) = \ell_t^y$, $y \in \mathbb{R}$, the function, in the space variable y, of the local times of Brownian motion at time t.

We shall use the notation:

$$\tilde{H}_t(a) = \lim_{\varepsilon \to 0} \int_0^t \frac{ds}{(B_s - a)} 1_{(|B_s-a| \geq \varepsilon)} \qquad (10.2)$$

More generally, we can define, for $\alpha < 3/2$:

$$\tilde{H}_t^{(\alpha)}(a) = \lim_{\varepsilon \to 0} \int_0^t \frac{ds}{(B_s - a)^{\tilde{\alpha}}} 1_{(|B_s-a| \geq \varepsilon)} \qquad (10.3)$$

with $x^{\tilde{\alpha}} \stackrel{\text{def}}{=} |x|^{\alpha} \mathrm{sgn}(x)$.

We shall simply note $\tilde{H}_t$ for $\tilde{H}_t(0)$, and $\tilde{H}_t^{(\alpha)}$ for $\tilde{H}_t^{(\alpha)}(0)$.

These processes (in the variable t) are quite natural examples of processes with zero energy, which have been studied, in particular, by Fukushima [45].

They also inherit a scaling property from Brownian motion, which partly explains why they possess some interesting distributional properties, when taken at certain random times, as will be proved in this chapter.

Moreover, the one-sided version of $\tilde{H}^{(\alpha)}$ plays an essential role in the representation of Bessel processes with dimension $d < 1$, as shown recently by Bertoin ([7], [8]). In fact, an important part of this chapter shall be devoted to the description of a new kind of excursion theory for Bessel processes with dimension $d < 1$, developed by Bertoin, and to some of its applications.

To conclude this introduction, a few words about the origin of such studies is certainly in order: to our knowledge, they may be traced back to Itô-Mc Kean ([50], Problem 1, p. 72) and Yamada's original papers ([94], [95], [96]).

10.1 Yamada's formulae

(10.1.1) To begin with, we remark that, if $(\ell_t^a; a \in \mathbb{R}, t \geq 0)$ denotes the family of Brownian local times, then, for a given $x \in \mathbb{R}$, and $\varepsilon > 0$, we have:

$$\int_{x-\varepsilon}^{x+\varepsilon} \frac{dy}{|y-x|^\gamma} |\ell_t^y - \ell_t^x| < \infty , \quad \text{as soon as: } \gamma < \tfrac{3}{2} ,$$

thanks to the following Hölder continuity property of Brownian local times:

$$\text{for } 0 < \eta < \tfrac{1}{2}, \quad \sup_{s \leq t} |\ell_s^a - \ell_s^b|(\omega) \leq C_{t,\eta}(\omega)|a - b|^{\frac{1}{2}-\eta} ,$$

for some (random) constant $C_{t,\eta}(\omega)$.

Consequently, the quantities: $(\tilde{H}_t^{(\beta)}(a); a \in \mathbb{R}, t \geq 0)$ are well-defined for any $\beta < \tfrac{3}{2}$.

Likewise, so are the quantities:

$$\text{p.v.} \int_0^t \frac{1_{(B_s-a>0)}ds}{(B_s-a)^{1+\alpha}} \overset{\text{def}}{=} \int_0^\infty \frac{db}{b^{1+\alpha}}(\ell_t^{a+b} - \ell_t^a)$$

and

$$\text{p.v.} \int_0^t \frac{ds}{|B_s-a|^{1+\alpha}} \overset{\text{def}}{=} \int_{-\infty}^\infty \frac{db}{|b|^{1+\alpha}}(\ell_t^{a+b} - \ell_t^a) \ ,$$

for $0 < \alpha < \frac{1}{2}$.

(10.1.2) The quantities we have just defined appear in fact as the zero quadratic variation parts in the canonical decompositions as Dirichlet processes of

$$(B_t - a)^{\widetilde{1-\alpha}}, \big((B_t - a)^+\big)^{1-\alpha} \quad \text{and} \quad |B_t - a|^{1-\alpha} \ , \quad \text{for } 0 < \alpha < \frac{1}{2} \ .$$

For simplicity, we shall take $a = 0$; then, we have the following formulae:

$$(B_t)^{\widetilde{1-\alpha}} = (1-\alpha) \int_0^t (B_s)^{-\tilde{\alpha}} dB_s + \frac{(1-\alpha)(-\alpha)}{2} \ \text{p.v.} \int_0^t \frac{ds}{B_s^{\widetilde{1+\alpha}}} \qquad (10.4)$$

$$(B_t^+)^{1-\alpha} = (1-\alpha) \int_0^t (B_s)^{-\alpha} 1_{(B_s>0)} dB_s + \frac{(1-\alpha)(-\alpha)}{2} \ \text{p.v.} \int_0^t \frac{1_{(B_s>0)}ds}{B_s^{1+\alpha}}$$
$$(10.5)$$

$$|B_t|^{1-\alpha} = (1-\alpha) \int_0^t |B_s|^{-\alpha} \text{sgn}(B_s) \, dB_s + \frac{(1-\alpha)(-\alpha)}{2} \ \text{p.v.} \int_0^t \frac{ds}{|B_s|^{1+\alpha}}$$
$$(10.6)$$

Exercise 10.1 In Revuz-Yor ([81], p. 230), the representation of the local time ℓ_t^y of Brownian motion, for fixed y, and fixed t, as an Itô stochastic integral, is given in the following explicit form:

$$\ell_t^y = \int_0^t ds \ g_s(y) - \frac{1}{\sqrt{2\pi}} \int_0^t \text{sgn}(B_s - y) q\left(\frac{B_s - y}{\sqrt{t-s}}\right) dB_s \ ,$$

where: $q(x) = 2 \int_{|x|}^\infty du \exp\left(-\frac{u^2}{2}\right)$, and $g_s(y) = \frac{1}{\sqrt{2\pi s}} \exp\left(-\frac{y^2}{2s}\right)$.

Derive from this formula the representation as an Itô integral of the different principal values we have just defined, in particular: $\int\limits_0^t \dfrac{ds}{B_s}$.

(10.1.3) We shall now transform formula (10.6) into a formula which gives the canonical decomposition, as a Dirichlet process, of a Bessel process $(R_t^{(\delta)}, t \geq 0)$, with dimension δ, such that: $0 < \delta < 1$. We first recall that a power of a Bessel process is another Bessel process time-changed; precisely, we have the formula:

$$q R_\nu^{1/q}(t) = R_{\nu q}\left(\int\limits_0^t \dfrac{ds}{R_\nu^{2/p}(s)}\right) \qquad (10.7)$$

where $(R_\mu(t), t \geq 0)$ denotes a Bessel process with index μ, and $\nu > -\frac{1}{q}$, $\frac{1}{p} + \frac{1}{q} = 1$ (see, e.g.: Revuz-Yor ([81], Proposition (1.11), p. 416); in fact, formula (10.7) was already presented and used in Chapter 9, as formula (9.2)). Applying this formula with $\nu = -\frac{1}{2}$ (so that $(R_\nu(t), t \geq 0)$ is a reflecting Brownian motion, and $R_{\nu q}(t) \equiv R^{(\delta)}(t), t \geq 0$), we obtain the following consequence of formula (10.6):

$$R_t \equiv R^{(\delta)}(t) = \beta_t + \dfrac{\delta - 1}{2} K_t \qquad (10.8)$$

where $(\beta_t, t \geq 0)$ is a Brownian motion, and:

$$K_t = \text{p.v.} \int\limits_0^t \dfrac{ds}{R_s} \overset{\text{def}}{=} \int\limits_0^\infty a^{\delta-2} da (L_t^a - L_t^0) \ ,$$

the family of local times $(L_t^a, a \geq 0)$ being defined with respect to the speed measure of $R^{(\delta)}$ as:

$$\int\limits_0^t ds \varphi(R_s) = \int\limits_0^\infty da \varphi(a) L_t^a a^{\delta-1}$$

for every Borel function $\varphi : \mathbb{R}_+ \to \mathbb{R}_+$.

10.2 A construction of stable processes, involving principal values of Brownian local times

(10.2.1)   Let $\alpha \in]-\infty, \frac{3}{2}[$. With the help of the scaling property of the process $(\tilde{H}_t^{(\alpha)}, t \geq 0)$, and using the inverse $\tau_t \equiv \inf\{u : \ell_u^0 > t\}$ of the Brownian local time $(\ell_u^0, u \geq 0)$, it is easy to construct symmetric stable processes from a 1-dimensional BM. Precisely, we have

Theorem 10.1 *Let $\alpha \in] - \infty, \frac{3}{2}[$. Then, the process $(\tilde{H}_{\tau_t}^{(\alpha)}, t \geq 0)$ is a symmetric stable process of index $\nu_\alpha = \dfrac{1}{2-\alpha}$; in particular, we have:*

$$E\left[\exp(i\lambda \tilde{H}_{\tau_t}^{(\alpha)})\right] = \exp(-t\, c_\alpha |\lambda|^{\nu_\alpha}) \qquad (\lambda \in \mathbb{R})$$

for some constant c_α.

Remarks:

1) As α varies from $-\infty$ to $\frac{3}{2}$ (excluded), ν_α varies from 0 to 2, with extreme values excluded; hence, with this construction, we can obtain all symmetric stable processes, except Brownian motion!

2) In the particular case $\alpha = 1$, $(\tilde{H}_{\tau_t}, t \geq 0)$ is a multiple of the standard Cauchy process. In fact, as we shall see with the next theorem, $\left(\dfrac{1}{\pi}\tilde{H}_{\tau_t}, t \geq 0\right)$ is a standard Cauchy process.

3) P. Fitzsimmons and R. Getoor [40] have extended the result concerning $\left(\tilde{H}_{\tau_t}, t \geq 0\right)$ to a large class of symmetric Lévy processes in place of the Brownian motion. They were also intrigued by the presence of the constant π. The computations of Fitzsimmons and Getoor have been simplified and generalized by Bertoin [9], using stochastic calculus and Feynman-Kac arguments.

(10.2.2)   It now seems natural to look for some relation between the results of Theorem 10.1 and a more classical construction of the stable symmetric processes, which may be obtained as time-changes of a Brownian motion by an independent unilateral stable process. More precisely, Spitzer [85] remarked that, if $(\gamma_u, u \geq 0)$ is another real-valued Brownian motion, which is independent of B, then:

$$(\gamma_{\tau_t}, t \geq 0) \quad \text{is a standard symmetric Cauchy process} \qquad (10.9)$$

Molchanov-Ostrovski [67] replaced $(\tau_t, t \geq 0)$ by any unilateral stable process to obtain all symmetric stable processes, except Brownian motion. J.F. Le Gall [59] presented yet another construction in the general case, which is closer to Spitzer's original idea, in that it involves complex Brownian motion.

In any case, coming back precisely to Theorem 10.1 (or, rather, to the second remark following it) and Spitzer's result (10.9), we see that $\left(\dfrac{1}{\pi}\tilde{H}_u, u \geq 0\right)$ and $(\gamma_u, u \geq 0)$, when restricted to the zero set of the Brownian motion $(B_v, v \geq 0)$, have the same law. Therefore, it now seems natural to consider their joint distribution for fixed time t.

Theorem 10.2 (We keep the previous notation concerning the independent Brownian motions B and γ).

For every $\lambda \in \mathbb{R}$, *and* $\theta \neq 0$, *we have:*

$$E\left[\exp i\left(\frac{\lambda}{\pi}\tilde{H}_{\tau_t} + \theta\gamma_{\tau_t}\right)\right] = E\left[\exp\left(i\frac{\lambda}{\pi}\tilde{H}_{\tau_t} - \frac{\theta^2}{2}\tau_t\right)\right] = \exp\left(-t\lambda\coth\left(\frac{\lambda}{\theta}\right)\right).$$

This formula is reminiscent of Lévy's stochastic area formula (2.7); it seems to call for some interpretation in terms of complex Brownian motion, which we shall attempt, with some partial success, in the next paragraph.

10.3 Distributions of principal values of Brownian local times, taken at an independent exponential time

We start again with the interesting case $\alpha = 1$. It will be fruitful to decompose the process $(\tilde{H}_t, t \geq 0)$ into the sum of:

$$\tilde{H}_t^- = \tilde{H}_{g_t} \quad \text{and} \quad \tilde{H}_t^+ = \tilde{H}_t - \tilde{H}_{g_t} , \quad \text{where } g_t = \sup\{s \leq t : B_s = 0\}$$

Theorem 10.3 *Let* T *denote a r.v. with values in* $\mathbb{R}_+$, *which is exponentially distributed, with parameter* $\frac{1}{2}$; *moreover,* T *is assumed to be independent of* B. *Then, we have the following:*

i) $\tilde{H}_T^-$ *and* $\tilde{H}_T^+$ *are independent;*

ii) *for every* $\lambda \in \mathbb{R}$,

$$E\left[\exp\left(i\frac{\lambda}{\pi}\tilde{H}_T^-\right)\right] = \frac{\text{th}(\lambda)}{\lambda} \quad and \quad E\left[\exp\left(i\frac{\lambda}{\pi}\tilde{H}_T^+\right)\right] = \frac{\lambda}{\text{sh}(\lambda)} \quad .$$

Therefore, we have:

$$E\left[\exp\left(i\frac{\lambda}{\pi}\tilde{H}_T\right)\right] = \frac{1}{\text{ch}(\lambda)} \tag{10.10}$$

iii) *In fact, formula (10.10) may be completed as follows:*

$$E\left[\exp\left(i\frac{\lambda}{\pi}\tilde{H}_T\right) \mid \ell_T^0 = t\right] = \frac{\lambda}{\text{sh}(\lambda)}\exp -t(\lambda\coth\lambda - 1) \quad . \tag{10.11}$$

10.4 Bertoin's excursion theory for BES(d), $0 < d < 1$

In this paragraph, $(R_t, t \geq 0)$ denotes a BES(d) process, with $0 < d < 1$, and $(K_t, t \geq 0)$ is the process with zero quadratic variation such that:

$$R_t = R_0 + B_t + (d-1)K_t \qquad (t \geq 0) \ ,$$

a decomposition we already encountered in paragraph 10.1, formula (10.8), with the factor $(\frac{1}{2})$ deleted.

Bertoin [8] proved that $(0,0)$ is regular for itself, with respect to the Markov process (R, K); hence, it admits a local time; such a local time $(\delta(t), t \geq 0)$ may be constructed explicitly from K as the limit of $2^{n(d-1)}d_n(t)$, where $d_n(t)$ denotes the number of downcrossings of K from 0 to -2^{-n} during the time-interval $[0, t]$.

Let $\sigma(t) = \inf\{s : \delta(s) > t\}$ be the right-continuous inverse of δ, and consider the Poisson point process: $e = (e^1, e^2)$ defined by:

$$e^1(t) = \{R_{\sigma(t-)+h}\,1_{(h \leq \sigma(t)-\sigma(t-))}; h \geq 0\}$$
$$e^2(t) = \{K_{\sigma(t-)+h}\,1_{(h \leq \sigma(t)-\sigma(t-))}; h \geq 0\}$$

Call **m** the (Itô) characteristic measure of this Poisson point process, which lives on Ω_0^{abs}, the set of continuous functions $\varepsilon : \mathbb{R}_+ \to \mathbb{R}_+ \times \mathbb{R}$, such that $\varepsilon(0) = (0,0)$, and ε is <u>absorbed</u> at $(0,0)$ after its first return $V(\varepsilon)$ to $(0,0)$.

For $\varepsilon \in \Omega_0^{abs}$, we define furthermore: $U(\varepsilon) = \inf \{t > 0 : \varepsilon^2(t) = 0\}$. We may now state Bertoin's description of **m**.

Theorem 10.4 *The σ-finite measure **m** is characterized by the following distributional properties:*

1) $\mathbf{m}(d\varepsilon)$ *a.s.,* $\left(\varepsilon^2(t), t \leq U\right)$ *takes values in* $\mathbb{R}_-$, *and* $\left(\varepsilon^2(t), U \leq t \leq V\right)$ *takes values in* $\mathbb{R}_+$;

2) $\mathbf{m}\left(\varepsilon^1(U) \in dx\right) = \dfrac{1 - d}{\Gamma(d)} x^{d-2} dx \quad (x > 0)$

3) *Conditionally (with respect to **m**) on* $\varepsilon^1(U) = x$, *the processes:*

$$\left(\varepsilon^1(U - h), -\varepsilon^2(U - h); h \leq U\right) \ \text{and} \ \left(\varepsilon^1(U + h), \varepsilon^2(U + h); h \leq V - U\right)$$

are independent, and have both the same distribution as:

$$\left(R_x(t), K_x(t); t \leq S_x\right),$$

where $(R_x(t), t \geq 0)$ *denotes a* $BES_x(d)$ *process, with canonical (Dirichlet) decomposition:*
$$R_x(t) = x + B_t + (d-1)K_x(t) \ ,$$
and $S_x = \inf \{t : K_x(t) = 0\}$.

Bertoin [8] deduced several distributional results from Theorem 10.4. In turn, we shall use Theorem 10.4 to characterize the law of

$$A_1^+ = \int\limits_0^1 ds \, 1_{(K_s > 0)} \ .$$

Recall that, from excursion theory, we have, for any continuous, increasing additive functional $(A_t, t \geq 0)$ of $X \equiv (R, K)$, which does not charge $\{s : R_s = K_s = 0\}$, the following formulae:

$$E\left[\int_0^\infty dt \exp-(\alpha t + A_t)\right] = \frac{\int \mathbf{m}(d\varepsilon) \int_0^V dt \exp-(\alpha t + A_t)}{\int \mathbf{m}(d\varepsilon)\,(1 - \exp-(\alpha V + A_V))}$$

$$(10.12)$$

$$E\left[\int_0^\infty dt \exp-(\alpha t + A_{g_t})\right] = \frac{\frac{1}{\alpha}\int \mathbf{m}(d\varepsilon)(1 - \exp(-\alpha V))}{\int \mathbf{m}(d\varepsilon)(1 - \exp-(\alpha V + A_V))}$$

We now apply these formulae with: $A_t = \beta A_t^+ + \gamma A_t^-$, where $A_t^- = t - A_t^+$; the quantities to be computed are:

$$h(\alpha, \beta, \gamma) \stackrel{\text{def}}{=} \int \mathbf{m}(d\varepsilon)\left(1 - \exp-(\alpha V + \beta A_V^+ + \gamma A_V^-)\right)$$
$$= \int \mathbf{m}(d\varepsilon)\left(1 - \exp-\{(\alpha + \gamma)U + (\alpha + \beta)(V - U)\}\right)$$

and

$$k(\alpha, \beta, \gamma)$$
$$\stackrel{\text{def}}{=} \int \mathbf{m}(d\varepsilon) \int_0^V dt \exp-(\alpha t + \beta A_t^+ + \gamma A_t^-)$$
$$= \int \mathbf{m}(d\varepsilon)\left\{\int_0^U dt \exp\left(-(\alpha + \gamma)t\right) + \int_U^V dt \exp-(\alpha t + \beta(t - U) + \gamma U)\right\}.$$

Hence, if we now define:

$$f(a, b) = \int \mathbf{m}(d\varepsilon)\left(1 - \exp-(aU + b(V - U))\right)$$

we obtain, with a little algebra: $h(\alpha, \beta, \gamma) = f(\alpha + \gamma, \alpha + \beta)$ and

$$k(\alpha, \beta, \gamma) = \frac{1}{\alpha + \beta}\left[\frac{(\beta - \gamma)f(\alpha + \gamma, 0)}{\alpha + \gamma} + f(\alpha + \gamma, \alpha + \beta)\right] \qquad (10.13)$$

We are now in a position to state the following

Theorem 10.5 *1) For every $t \geq 0$, one has:*

$$E\left[\exp-(aA_{\sigma(t)}^- + bA_{\sigma(t)}^+)\right] = \exp-tf(a, b) \ ,$$

where: $f(a, b) = \left(\sqrt{2a} + \sqrt{2b}\right)^{1-d}.$

2) The distributions of the variable A_1^+ and of the pair $(A_{g_1}^+, A_{g_1}^-)$ are characterized by the formulae:

$$E\left[\frac{1}{1 + \beta A_1^+}\right] = \frac{\beta + \left(1 + \sqrt{1 + \beta}\right)^{1-d}}{(1 + \beta)\left(1 + \sqrt{1 + \beta}\right)^{1-d}}$$

$$\hspace{8cm} (10.14)$$

$$E\left[\frac{1}{1 + \beta A_{g_1}^+ + \gamma A_{g_1}^-}\right] = \left(\frac{2}{\sqrt{1 + \beta} + \sqrt{1 + \gamma}}\right)^{1-d}$$

In particular, g_1 is distributed as:

$$Z_{\frac{1-d}{2}, \frac{1+d}{2}}, \quad \text{a beta variable with parameters } \left(\tfrac{1-d}{2}, \tfrac{1+d}{2}\right).$$

PROOF: 1) Bertoin ([8], Theorem 4.2) proved that if $(\lambda_t^a; a \in \mathbb{R})$ denotes the family of occupation densities of K, which are defined by:

$$\int_0^t ds\, f(K_s) = \int_{-\infty}^\infty da\, f(a)\lambda_t^a \ ,$$

then, conditionally on $\lambda_{\sigma(t)}^0 = x$, the processes $(\lambda_{\sigma(t)}^a, a \geq 0)$ and $(\lambda_{\sigma(t)}^{-a}, a \geq 0)$ are two independent $\mathrm{BESQ}_x(0)$ processes.

Furthermore, the law of $\lambda_{\sigma(t)}^0$ is characterized by:

$$E\left[\exp\left(-\frac{k}{2}\lambda_{\sigma(t)}^0\right)\right] = \exp(-tk^{1-d}) \qquad (k \geq 0) \ .$$

Using this result, we obtain:

$$E\left[\exp - \left(a A_{\sigma(t)}^- + b A_{\sigma(t)}^+\right)\right] = E\left[\exp -\frac{\lambda_{\sigma(t)}^0}{2}\left(\sqrt{2a} + \sqrt{2b}\right)\right]$$

$$= \exp -t\left(\sqrt{2a} + \sqrt{2b}\right)^{1-d}.$$

2) It follows from formulae (10.12) and (10.13) that:

$$E\left[\int_0^\infty dt\, \exp - (\alpha t + \beta A_t^+)\right] = \frac{k(\alpha, \beta, 0)}{h(\alpha, \beta, 0)} = \frac{\beta f(\alpha, 0) + \alpha f(\alpha, \alpha + \beta)}{\alpha(\alpha + \beta)f(\alpha, \alpha + \beta)}$$

and

$$E\left[\int_0^\infty dt \exp -(\alpha t + \beta A_{g_t}^+ + \gamma A_{g_t}^-)\right] = \frac{h(\alpha,0,0)}{\alpha\, h(\alpha,\beta,\gamma)} = \frac{f(\alpha,\alpha)}{\alpha f(\alpha+\gamma,\alpha+\beta)}.$$

Now, the expectations on the left-hand sides of these equalities are respectively equal, using a scaling argument, to:

$$E\left[\frac{1}{\alpha + \beta A_1^+}\right] \quad \text{and} \quad E\left[\frac{1}{\alpha + \beta A_{g_1}^+ + \gamma A_{g_1}^-}\right].$$

The proof is ended by replacing $f(a,b)$ by $\left(\sqrt{2a} + \sqrt{2b}\right)^{1-d}$ in the above equalities. $\square$

Remark: It may be interesting to compare formula (10.14) with yet another distributional result:

$$\text{for fixed } t, \qquad \frac{A_{\sigma(t)}^+}{\sigma(t)} \stackrel{(\text{law})}{=} \frac{A_{\sigma(t)}^-}{\sigma(t)} \stackrel{(\text{law})}{=} Z_{\frac{1}{2},\frac{1}{2}}, \qquad (10.15)$$

i.e.: both ratios are arc sine distributed.

This follows immediately from the description of the law of $(\lambda_{\sigma(t)}^a, a \in \mathbb{R})$ already used in the above proof.

Comments on Chapter 10

The contents of this chapter consist mainly of results relating principal values for Bessel processes with small dimension, and their excursion theory, as derived by Bertoin [8]. For a further discussion by Bertoin, see [10].

A more complete exposition of results pertaining to principal values of local times is given in Yamada [97], and also in the second half of the Monograph [103], which centers around Alili's study of:

$$\text{p.v.} \int_0^t ds \, \coth(\lambda B_s) \, ,$$

and the, up to now little understood , striking identity:

$$\lambda^2\left(\left(\int_0^1 ds \, \coth(\lambda r_s)\right)^2 - 1\right) \stackrel{(\text{law})}{=} \left(\int_0^1 \frac{ds}{r_s}\right)^2$$

where $(r_s, s \le 1)$ denotes the standard 3-dimensional Bessel bridge, and $\lambda \in \mathbb{R}$ (thus, the law of the left-hand side does not depend on λ).

More studies of functionals of $(r_s, s \leq 1)$, including $\int_0^1 ds \, \exp(\pm \lambda r_s)$ are also found in C. Donati-Martin and M. Yor [34].

Chapter 11

Probabilistic representations of the Riemann zeta function and some generalisations related to Bessel processes

To begin with, it may be wise to state immediately that the aim[1] of this chapter is not to discuss Riemann's hypothesis!, but, much more modestly, to present some of the (well-known) relations between heat equation, zeta function, theta functions and Brownian motion.

11.1 The Riemann zeta function and the 3-dimensional Bessel process

(11.1.1) The Riemann zeta function is defined by:

$$\zeta(s) = \sum_{n=1}^{\infty} \frac{1}{n^s}, \qquad \text{for } s \in \mathbb{C}, \text{ Re}(s) > 1.$$

It extends analytically to the entire complex plane $\mathbb{C}$, as a meromorphic function with a unique pole at $s = 1$.

An essential property of ζ is that it satisfies the functional equation:

$$\xi(s) = \xi(1 - s) \tag{11.1}$$

---

[*] Researches linking the Riemann Zeta function and random matrix theory, in particular: "the Keating-Snaith philosophy", which is closely related to the Lindelöf hypothesis, are beyond the scope of this book. However see e.g. the Mezzadri-Snaith volume [66]

where:
$$\xi(s) \stackrel{\text{def}}{=} \frac{s(s-1)}{2} \, \Gamma\left(\frac{s}{2}\right) \, \pi^{-s/2} \zeta(s) \,. \tag{11.2}$$

We recall that the classical gamma function, which is defined by:

$$\Gamma(s) = \int_0^\infty dt \, t^{s-1} e^{-t} \,, \qquad\qquad \text{for Re}(s) > 0 \,,$$

extends analytically to $\mathbb{C}$ as a meromorphic function with simple poles at $0, -1, -2, \ldots, -m, \ldots$, thanks to the relation:

$$\Gamma(1 + s) = s \, \Gamma(s) \,.$$

(11.1.2) The functional equation (11.1) may be understood as a symmetry property of the distribution of the r.v.:

$$\mathcal{N} \stackrel{\text{def}}{=} \frac{\pi}{2} \, \mathbf{T}_{(2)} \,, \quad \text{where:} \; \mathbf{T}_{(2)} \stackrel{\text{def}}{=} T_1^{(3)} + \tilde{T}_1^{(3)} \,,$$

with $T_1^{(3)}$ and $\tilde{T}_1^{(3)}$ two independent copies of the first hitting time of 1 by a BES(3) process starting from 0.

Indeed, one has:
$$2\,\xi(2\,s) = E[\mathcal{N}^s] \tag{11.3}$$

Hence, if we assume that the functional equation (11.1) holds, we deduce from (11.3) that $\mathcal{N}$ satisfies:

$$E[\mathcal{N}^s] = E[\mathcal{N}^{(1/2)-s}] \,, \qquad \text{for any } s \in \mathbb{C} \,,$$

or, equivalently: for any Borel function $f : \mathbb{R}_+ \to \mathbb{R}_+$,

$$E[f(\mathcal{N})] = E\left[f\left(\frac{1}{\mathcal{N}}\right) \sqrt{\mathcal{N}}\right] \,. \tag{11.4}$$

In paragraphs 11.2 and 11.3, an explanation of this symmetry property of $\mathcal{N}$ is given.

(11.1.3) For the moment, we give a proof of (11.4), hence of (11.1), as a consequence of Jacobi's identity for the theta function:

$$\Theta\left(\frac{1}{t}\right) = \sqrt{t} \; \Theta(t) \,, \quad \text{where } \Theta(t) \equiv \sum_{n=-\infty}^{\infty} e^{-\pi n^2 t} \,. \tag{11.5}$$

Indeed, the density of $\mathcal{N}$, which we denote by $\varphi(t)$, satisfies:

$$\varphi(t) = 2t\,\Theta''(t) + 3\,\Theta'(t)\,,$$

and it is easily deduced from this identity that:

$$\varphi\left(\frac{1}{t}\right) = t^{5/2}\,\varphi(t) \tag{11.6}$$

which is equivalent to (11.4).

The following exercise should help to understand better the deep connections which exist between the Riemann zeta function and the distribution of $T_1^{(3)}$ (and its powers of convolution).

Exercise 11.1 Let $k > 0$, and let $\mathbf{T}_{(k)}$ denote an $\mathbb{R}_+$-valued r.v. such that

$$E\left[\exp\left(-\frac{\lambda^2}{2}\,\mathbf{T}_{(k)}\right)\right] = \left(\frac{\lambda}{\mathrm{sh}\lambda}\right)^k$$

(such a variable exists, thanks to the infinite divisibility of $\mathbf{T}_{(1)}$; from formula (2.6), $\mathbf{T}_{(k)}$ may be represented as: $\int_0^1 ds\,\rho_{(k)}^2(s)$, where $(\rho_{(k)}(s),\ s \le 1)$ denotes here the $(2k)$-dimensional Bessel bridge).

1. Prove that, for any $m > 0$, one has:

$$\Gamma(m)\,E\left[\frac{1}{(\mathbf{T}_{(k)})^m}\right] = \frac{1}{2^{m-k-1}}\int_0^\infty d\lambda\,\lambda^{k+2m-1}\frac{e^{-\lambda k}}{(1-e^{-2\lambda})^k}$$

2. Assume k is an integer, $k \ge 1$. Recall that $\dfrac{1}{1-x} = \displaystyle\sum_{n=0}^\infty x^n$, $(x < 1)$ and, for $k \ge 2$:

$$\frac{(k-1)!}{(1-x)^k} = \sum_{n=k-1}^\infty n(n-1)\cdots(n-(k-2))x^{n-(k-1)}$$

More generally, for any $k > 0$, we have

$$\frac{1}{(1-x)^k} = \sum_{p=0}^\infty \alpha_p^{(k)}\,x^p\,, \quad \text{with } \alpha_p^{(k)} = \frac{\Gamma(k+p)}{\Gamma(k)\,\Gamma(p+1)}\,.$$

Deduce, from the first question, that:

$$\Gamma(m)\, E\left[\frac{1}{(\mathbf{T}_{(k)})^m}\right] = \frac{\Gamma(k+2m)}{2^{m-k-1}} \sum_{p=0}^{\infty} \frac{\alpha_p^{(k)}}{(k+2p)^{k+2m}}$$

3. Show the following formulae for $E\left[\dfrac{1}{(\mathbf{T}_{(k)})^m}\right]$, with $k = 1, 2, 3, 4$, in terms of Γ and ζ.

$$E\left[\frac{1}{(\mathbf{T}_{(1)})^m}\right] = \frac{\Gamma(2m+1)}{(2^{m-2}\,\Gamma(m))} \left(\sum_{n=0}^{\infty} \frac{1}{(2n+1)^{2m+1}}\right)$$

$$= \frac{\Gamma(2m+1)}{(2^{m-2}\,\Gamma(m))} \left(1 - \frac{1}{2^{2m+1}}\right) \zeta(2m+1)\,.$$

$$E\left[\frac{1}{(\mathbf{T}_{(2)})^m}\right] = \frac{\Gamma(2m+2)}{(2^{3m-1}\Gamma(m))} \zeta(2m+1)\,.$$

$$E\left[\frac{1}{(\mathbf{T}_{(3)})^m}\right] = \frac{\Gamma(2m+3)}{2^{m-1}\,\Gamma(m)} \left\{ \left(1 - \frac{1}{2^{2m+1}}\right) \cdots \right.$$

$$\left. \cdots\zeta(2m+1) - \left(1 - \frac{1}{2^{2m+3}}\right) \zeta(2m+3)\right\}\,.$$

$$E\left[\frac{1}{(\mathbf{T}_{(4)})^m}\right] = \frac{\Gamma(2m+4)}{3\cdot 2^{3m-2}\,\Gamma(m)} \left\{\zeta(2m+1) - \zeta(2m+3)\right\}\,.$$

Prove that, for any integer $k \geq 1$, it is possible to express $E\left[\dfrac{1}{(\mathbf{T}_{(k)})^m}\right]$ in terms of the Γ and ζ functions.

4. Deduce, from the comparison of the expressions of $E\left[\dfrac{1}{(\mathbf{T}_{(1)})^m}\right]$ and $E\left[\dfrac{1}{(\mathbf{T}_{(2)})^m}\right]$ that:

$$(*) \qquad\qquad \frac{U^2}{\mathbf{T}_{(2)}} \stackrel{\text{(law)}}{=} \frac{Y^2}{\mathbf{T}_{(1)}} \left(\stackrel{\text{(law)}}{=} Y^2 (\sup_{u\leq 1} R_u^{(3)})^2\right),$$

where U denotes a uniform r.v., independent of $\mathbf{T}_{(2)}$, and Y a discrete r.v. independent of $\mathbf{T}_{(1)}$ and such that: $P\left(Y = \dfrac{1}{2^p}\right) = \dfrac{1}{2^p}, \ (p = 1, 2, \ldots).$

11.2 The right hand side of (11.4), and the agreement formulae between laws of Bessel processes and Bessel bridges

(11.2.1) Using (Brownian) excursion theory, we will show below that, for every Borel function $f : \mathbb{R}_+ \to \mathbb{R}_+$, one has:

$$E[f(m_e^2)] = \sqrt{\frac{\pi}{2}} E\left[f\left(\frac{1}{\mathbf{T}_{(2)}} \right) \sqrt{\mathbf{T}_{(2)}} \right] \tag{11.7}$$

where $(e(u), \ u \leq 1)$ denotes the *normalized Brownian excursion*, which is distributed as the 3-*dimensional standard Bessel bridge*, and $m_e \overset{\text{def}}{=} \sup_{u \leq 1} e(u)$.

Assuming (11.7) holds, it will remain, in order to finish the proof of (11.4) to show:

$$m_e^2 \overset{\text{(law)}}{=} \frac{\pi^2}{4} \mathbf{T}_{(2)} \tag{11.8}$$

which will be undertaken in paragraph 11.3.

(11.2.2) The identity (11.7) will appear below as a particular consequence of the following agreement formulae which are now presented as relationships, for any dimension $d > 0$, between the law of the standard d-dimensional Bessel bridge on one hand, and, on the other hand, of the law of two d-dimensional Bessel processes put back to back. Here is this relationship:

Theorem 11.1 *Let $d > 0$, and define $\mu = \frac{d}{2} - 1$.*

Consider $(R_u, \ u \geq 0)$ and $(R'_u, \ u \geq 0)$ two independent $\mathrm{BES}^\mu \equiv \mathrm{BES}(d)$-processes starting from 0; denote[2] by σ_μ and σ'_μ their respective first hitting times of 1.

Let

$$\rho_u = \begin{cases} R_u & , \text{ if } u \leq \sigma_\mu \\ R'_{\sigma_\mu + \sigma'_\mu - u} & , \text{ if } \sigma_\mu \leq u \leq \sigma_\mu + \sigma'_\mu, \end{cases}$$

and

$$\tilde{\rho}_v = \frac{1}{\sqrt{\sigma_\mu + \sigma'_\mu}} \rho_{v(\sigma_\mu + \sigma'_\mu)}, \ v \leq 1.$$

Then, if $(r_v, \ v \leq 1)$ denotes the standard Bessel bridge with dimension d, we have, for every measurable functional $F : C([0, 1], \mathbb{R}_+) \to \mathbb{R}_+$:

---

[*] Thus, σ_μ is another (sometimes more convenient) notation for $T_.^{\cdot d \cdot}$.

$$E[F(r_v, v \leq 1)] = 2^\mu\, \Gamma(\mu+1)\, E[F(\tilde\rho_v, v \leq 1)(\sigma_\mu + \sigma'_\mu)^\mu]. \qquad (11.9)$$

We now remark that the identity (11.7) follows from the identity (11.10) below, in the case $\mu = 1/2$.

Corollary 11.1.1 *Let m_μ be the supremum of the standard Bessel bridge with dimension $d = 2(1+\mu)$, and let s_μ be the unique time at which this supremum is attained. Then, we have, for every Borel function $f : \mathbb{R}_+^2 \to \mathbb{R}_+$,*

$$E[f(m_\mu^2, s_\mu)] = 2^\mu\, \Gamma(\mu+1)\, E\left[f\left(\frac{1}{\sigma_\mu + \sigma'_\mu}, \frac{\sigma_\mu}{\sigma_\mu + \sigma'_\mu}\right)(\sigma_\mu + \sigma'_\mu)^\mu\right].$$
$$(11.10)$$

PROOF: This is immediate from the identity (11.9) above, since m_μ^2, resp. s_μ, considered on the left hand side of (11.9), corresponds to $1/(\sigma_\mu + \sigma'_\mu)$, resp. $\sigma_\mu/(\sigma_\mu + \sigma'_\mu)$, considered on the right hand side of (11.9). □

It should be noted, although this is a digression from our main theme, that, in the particular case $\mu = 0$ (or $d = 2$), Theorem 11.1 yields a remarkable identity in law.

Theorem 11.2 *We use the same notation as in Theorem 11.1, but now $d = 2$.*

Then, we have: $\qquad\qquad (r_v,\ v \leq 1) \stackrel{(\text{law})}{=} (\tilde\rho_v,\ v \leq 1).$

Corollary 11.2.1 *We use the same notations as in Corollary 11.1.1, but now $\mu = 0$ (or, $d = 2$). Then, we have:*

$$(m_0^2, s_0) \stackrel{(\text{law})}{=} \left(\frac{1}{\sigma_0 + \sigma'_0}, \frac{\sigma_0}{\sigma_0 + \sigma'_0}\right) \qquad (11.11)$$

and in particular:

$$\frac{s_0}{m_0^2} \stackrel{(\text{law})}{=} \sigma_0. \qquad (11.12)$$

(11.2.3) **A family of excursion measures** We now give a proof of Theorem 11.1, for $\mu > 0$, which relies upon two different descriptions, both due to D. Williams, of a σ−finite measure $\mathbf{n}_\mu$ already considered by Pitman-Yor ([73], p.436-440) and Biane-Yor ([17], paragraph (3.2)). $\mathbf{n}_\mu$ is defined on the canonical space $C(\mathbb{R}_+, \mathbb{R}_+)$, and is carried by the space Ω^{abs} of the trajectories ω, such that $\omega(0) = 0$ and ω is <u>absorbed</u> at 0 at the first (strictly positive) instant it reaches 0 again. $\mathbf{n}_\mu$ may be characterized by either of the following descriptions. For these descriptions, we shall use the notation:

$$e_u(\omega) = \omega(u) \ ; \quad V(\omega) = \inf\{u > 0 : e_u(\omega) = 0\} \ ; \quad M(\omega) = \sup_u e_u(\omega) \ .$$

First description of $\mathbf{n}_\mu$

(i) *The distribution of M under $\mathbf{n}_\mu$ is given by:*

$$\mathbf{n}_\mu(M \geq x) = x^{-2\mu} \qquad (x > 0) \ .$$

(ii) *For every $x > 0$, conditionally on $M = x$, this maximum M is attained at a unique time R $(0 < R < V$, a.s.$)$, and the two processes $(e_u, u \leq R)$ and $(e_{V-u}, u \leq V - R)$ are two independent BES_0^μ processes, stopped at the first time they reach level x.*

Second description of $\mathbf{n}_\mu$

(i') *The distribution of V under $\mathbf{n}_\mu$ is given by:*

$$\mathbf{n}_\mu(V \in dv) = \frac{\alpha_\mu dv}{v^{\mu+1}} \ , \quad \text{where } \alpha_\mu = \frac{1}{2^\mu \Gamma(\mu)}$$

(ii') *For every $v \in]0, \infty[$, conditionally on $V = v$, the process $(e_u, u \leq v)$ is a Bessel bridge of index μ, during the time interval $[0, v]$, starting and ending at 0.*

11.3 A discussion of the identity (11.8)

(11.3.1) The identity: $\qquad\qquad m_e^2 \overset{\text{(law)}}{=} \frac{\pi^2}{4} \mathbf{T}_{(2)} \qquad\qquad$ (11.8)

is reminiscent of the very well-known Kolmogorov-Smirnov identity:

$$m_b^2 \stackrel{\text{def}}{=} \sup_{u \leq 1}(b(u))^2 \stackrel{\text{(law)}}{=} \frac{\pi^2}{4} T_1^{(3)} \left(\stackrel{\text{(law)}}{=} T_{\pi/2}^{(3)} \right) \qquad (11.13)$$

where $(b(u), u \leq 1)$ denotes here the standard 1-dimensional Brownian bridge.

No satisfactory explanation has, until now, been given for the factor $(\pi/2)^2$ in either formula (11.8) or (11.13), but, putting them together, Chung [27] pointed out the puzzling identity in law:

$$m_e^2 \stackrel{\text{(law)}}{=} m_b^2 + m_{\tilde{b}}^2 \qquad (11.14)$$

where, on the right-hand side of (11.14), b and $\tilde{b}$ are two independent 1-dimensional Brownian bridges.

It follows from Vervaat's representation of the normalized Brownian excursion $(e(t), \ t \leq 1)$ (see Vervaat [89], and also Biane [12]), i.e.:

the process $\tilde{e}(t) \stackrel{\text{def}}{=} b((\rho+t) \,[\text{mod } 1]) - b(\rho), \ t \leq 1$, where ρ is the unique time at which b attains its minimum, is a normalized Brownian excursion, that:

$$m_e \stackrel{\text{(law)}}{=} \sup_{u \leq 1} b(u) - \inf_{u \leq 1} b(u) \,,$$

and, therefore, the identity (11.14) may be written as:

$$(\sup_{u \leq 1} b(u) - \inf_{u \leq 1} b(u))^2 \stackrel{\text{(law)}}{=} m_b^2 + m_{\tilde{b}}^2 \,. \qquad (11.15)$$

No pathwise explanation of the identities (11.14) or (11.15) has been found, and the explicit computation of the joint law of $(\sup_{u \leq 1} b(u), \inf_{u \leq 1} b(u))$ presented below in (11.3.2) rules out the possibility that (11.15) might be explained by the independence (which does not hold) of $\{(\sup_{u \leq 1} b(u) - \inf_{u \leq 1} b(u))^2 - m_b^2\}$ and m_b^2. To conclude with this series of identities, we use the well-known representation of brownian motion $(B_t, \ t \geq 0)$ in terms of the Brownian bridge $(b(u), \ u \leq 1)$:

$$B_t = (1+t)b \left(\frac{t}{1+t} \right), \quad t \geq 0,$$

from which it is easily deduced (see, e.g., Revuz-Yor [81], Exercise (3.10), p. 37) that:

$$\sup_{t \geq 0}(|B_t| - t) \stackrel{\text{(law)}}{=} \sup_{u \leq 1}(b(u))^2 \,.$$

Hence, we may write the identity in law (11.14) in the equivalent form:

$$\sup_{u \geq 0}(R(u) - u) \stackrel{(\text{law})}{=} \sup_{u \geq 0}(|B_u| - u) + \sup_{u \geq 0}(|\tilde{B}_u| - u)$$

$$\stackrel{(\text{law})}{=} \sup_{t \geq 0}(B_t^+ - \int_0^t ds\ 1_{(B_s \geq 0)}) + \sup_{t \geq 0}(B_t^- - \int_0^t ds\ 1_{(B_s \leq 0)})$$

(11.16)

where B and $\tilde{B}$ denote two independent Brownian motions, and $(R(u),\ u \geq 0)$ is a 3-dimensional Bessel process. (The last identity in law in (11.16) is left to the reader as an exercise; *Hint*: Use the representation of $B^{\pm}$ in terms of reflecting BM, given in Chapter 4, Paragraph 4.1).

(11.3.2) From the theory of Brownian excursions, the joint law of (s_b^+, s_b^-, ℓ_b), where: $s_b^+ = \sup_{u \leq 1} b(u)$, $s_b^- = -\inf_{u \leq 1} b(u)$, and ℓ_b is the local time at level 0 of the standard Brownian bridge $(b(u),\ u \leq 1)$ may be characterized as follows:

$$P(|G|s_b^+ \leq x; |G|s_b^- \leq y; |G|\ell_b \in d\lambda) = \exp(-\frac{\lambda}{2}(\coth x + \coth y))\ d\lambda \quad (11.17)$$

where G denotes a gaussian variable, centered, with variance 1, which is independent of b; consequently, one obtains, after integrating with respect to λ:

$$P(|G|s_b^+ \leq x; |G|s_b^- \leq y) = \frac{2}{\coth x + \coth y},$$

and it is now easy to deduce from this identity, together with the obvious equality:

$$m_b = \max(s_b^+, s_b^-),\ \text{that:}$$

$$E\left[\exp\left(-\frac{\alpha^2}{2}m_b^2\right)\right] = \frac{\left(\frac{\pi\alpha}{2}\right)}{\text{sh}\left(\frac{\pi\alpha}{2}\right)}$$

and

$$E\left[\exp\left(-\frac{\alpha^2}{2}(s_b^+ + s_b^-)^2\right)\right] = \left(\frac{\left(\frac{\pi\alpha}{2}\right)}{\text{sh}\left(\frac{\pi\alpha}{2}\right)}\right)^2.$$

This proves both identities (11.13) and (11.15) (and as we remarked above, (11.15) is equivalent to (11.14)).

We now remark, as an exercise, that the identity in law (11.15) may be translated into an identity in law between independent exponential and Bernoulli variables, the understanding of which does not seem obvious.

Exercise 11.2 (We keep the notation used in formula (11.17).)

1. Prove that:

$$|G| \, (\ell_b, 2\, s_b^+, 2\, s_b^-) \overset{\text{(law)}}{=} \left(T, \log(1 + \frac{T}{T'}), \log(1 + \frac{T}{T''}) \right)$$

and

$$|G| \, (\ell_b, 2m_b) \overset{\text{(law)}}{=} \left(T, \log(1 + \frac{2T}{T_*}) \right) ,$$

where (T, T', T''), resp. (T, T_*), are three, respectively two, independent exponential variables with parameter 1.

2. Prove that the identity in law (11.15) is equivalent to:

$$\left\{ \left(1 + \frac{T}{T'}\right) \left(1 + \frac{T}{T''}\right) \right\}^{\epsilon} \overset{\text{(law)}}{=} \left(1 + \frac{2T_1}{T'}\right)^{\epsilon'} \left(1 + \frac{2T_2}{T''}\right)^{\epsilon''}$$

where, on either side, the T's indicate independent exponential variables, which are also independent of the i.i.d. Bernoulli variables ϵ, ϵ' and ϵ'' $(P(\epsilon = \pm 1) = 1/2)$.

Here is now a proof of the identity (11.17).

Recall that the standard Brownian bridge $(b(u), \ u \le 1)$ may be represented as

$$\left(b(u) \equiv \frac{1}{\sqrt{g_t}} B_{ug_t}; \ u \le 1 \right) ,$$

where $g_t = \sup\{s < t; \ B_s = 0\}$.

Moreover, from excursion theory, we obtain the following equalities:

$$P(\ell_T \in d\lambda) = \exp(-\lambda)\, d\lambda,$$

and for any measurable functional $F : C([0, \infty), \mathbb{R}) \to \mathbb{R}_+$,

$$E[F(B_u, \ u \le g_T) \mid \ell_T = \lambda] = \exp(\lambda) E[F(B_u, \ u \le \tau_\lambda) \exp(-\frac{\tau_\lambda}{2})] \quad (11.18)$$

where $(\tau_\lambda, \ \lambda \ge 0)$ denotes the right continuous inverse of $(\ell_t, t \ge 0)$ and T denotes here an exponential time with parameter $1/2$, which is assumed to be independent of B.

Thanks to the first description of $\mathbf{n_{1/2}}$, which is given in 11.2, the following formula is easily obtained:

$$E\left[S^+_{\tau_\lambda} \le x; \; S^-_{\tau_\lambda} \le y; \; \exp\left(-\frac{\tau_\lambda}{2}\right)\right] = \exp\left(-\frac{\lambda}{2}(\coth x + \coth y)\right) \quad (11.19)$$

Furthermore, we remark that, by scaling

(i) $\quad \sqrt{g_T} \overset{\text{(law)}}{=} |G|$, and (ii) $\quad (S^+_{g_T}, S^-_{g_T}.\ell_T) \overset{\text{(law)}}{=} |G| \, (s^+_b, s^-_b, \ell_b)$,

where we have used the notation introduced at the beginning of this sub-paragraph 11.3. Now, in order to obtain formula (11.17), it remains to put together (i) and (ii) on one hand, and (11.18) and (11.19) on the other hand.

11.4 A strengthening of Knight's identity, and its relation to the Riemann zeta function

(11.4.1) In Chapter 9 of these Notes, we have given a proof, and some extensions of Knight's identity:

$$\text{for } \alpha \in \mathbb{R}, \qquad E\left[\exp\left(-\frac{\alpha^2}{2}\frac{\tau}{M^2_\tau}\right)\right] = \frac{2\alpha}{\text{sh}(2\,\alpha)} \qquad (11.20)$$

where, to simplify notations, we write τ instead of τ_1.

This identity (11.20) may be presented in the equivalent form:

$$\frac{\tau}{M^2_\tau} \overset{\text{(law)}}{=} T^{(3)}_2 \quad (:= \inf\{u; R_u = 2\}). \qquad (11.21)$$

We now remark that the identity (11.20), or (11.21), may be strengthened as follows.

Theorem 11.3 (Pitman-Yor [79]) *Define* $X = \dfrac{S^+_\tau}{S^+_\tau + S^-_\tau}$,

where $S^+_t = \sup\limits_{s \le t} B_s$, $S^-_t = -\inf\limits_{s \le t} B_s$.

Then, X *is uniformly distributed on* $[0,1]$, *and independent of* $\dfrac{\tau}{(S^+_\tau + S^-_\tau)^2}$,

which is distributed as $\mathbf{T}_{(2)} \overset{\text{(law)}}{=} T^{(3)}_1 + \tilde{T}^{(3)}_1$.

Equivalently, one has:

$$E\left[\exp\left(-\frac{\alpha^2}{2}\frac{\tau}{(S_\tau^+ + S_\tau^-)^2}\right)\right] = \left(\frac{\alpha}{\mathrm{sh}\alpha}\right)^2.$$

Theorem 11.3 constitutes indeed a strengthening of Knight's identity (11.20), since we can write:

$$\frac{\tau}{M_\tau^2} = \frac{\tau}{(S_\tau^+ + S_\tau^-)^2}\left(\max\left(X, 1 - X\right)\right)^{-2}$$

and it is easily shown that:

$$T_2^{(3)} \overset{(\mathrm{law})}{=} \mathbf{T}_{(2)}\left(\max\left(X, 1 - X\right)\right)^{-2},$$

where, on the right-hand side, $\mathbf{T}_{(2)}$ and X are assumed to be independent.

Exercise 11.3 1. Prove that, if X is uniformly distributed on $[0, 1]$, then $V = \max\left(X, 1 - X\right)$ is uniformly distributed on $[1/2, 1]$.

2. Prove that the identity in law: $T_2^{(3)} \overset{(\mathrm{law})}{=} \dfrac{\mathbf{T}_{(2)}}{V^2}$ we just encountered above

agrees with the identity in law $(*)$ $\dfrac{U^2}{\mathbf{T}_{(2)}} \overset{(\mathrm{law})}{=} \dfrac{Y^2}{T_1^{(3)}}$ derived in question 3

of Exercise 11.1.

Hint: Remark that $U \overset{(\mathrm{law})}{=} (2Y)\, V$, where, on the right hand side Y and V are independent.

A simple proof of Theorem 11.3 may be deduced from the identity (11.19), once we use the scaling property of BM to write the left-hand side of (11.19) as:

$$P\left(\lambda S_\tau^+ \le x; \lambda S_\tau^- \le y; \exp\left(-\frac{\lambda^2}{2}\tau\right)\right).$$

However, a more complete explanation of Theorem 11.3 may be given, in terms of a Vervaat-type theorem for the pseudo-bridge

$$\left(\frac{1}{\sqrt{\tau}}B_{u\tau}; u \le 1\right).$$

Theorem 11.4 ([79]; We keep the above notation)

Let ρ be the (unique) instant in the interval $[0, \tau]$ at which $(B_u; u \leq \tau)$ attains its minimum. Define the process $\tilde{B}$ as

$$(\tilde{B}(t); t \leq \tau) := (B((\rho + t)[mod\ \tau]) - B(\rho); t \leq \tau)$$

Then, denoting by $(e(u); u \leq 1)$ the normalized Brownian excursion, we have:

$$E\left[F\left(\frac{1}{\sqrt{\tau}}\tilde{B}(u\tau); u \leq 1\right)\right] = \sqrt{\frac{2}{\pi}} E\left[m_e F(e(u); u \leq 1)\right]$$

for any measurable $F : C([0, 1]; \mathbb{R}_+) \to \mathbb{R}_+$.

(11.4.2) The above strengthening of Knight's identity enables us to present now a very concise discussion of the identity in law (11.4), which we write in the equivalent form:

$$E[f(\mathbf{T}_{(2)})] = E\left[f\left(\frac{1}{(\pi^2/4)\,\mathbf{T}_{(2)}}\right)\sqrt{\frac{\pi}{2}}\,\mathbf{T}_{(2)}\right]. \qquad (11.22)$$

Indeed, the left-hand side of (11.22) is, from Theorem 11.3, equal to

$$E\left[f\left(\frac{\tau}{(S_\tau^+ + S_\tau^-)^2}\right)\right],$$

but, now from Theorem 11.4, this expression is also equal to:

$$\sqrt{\frac{2}{\pi}}\,E\left[f\left(\frac{1}{(s_b^+ + s_b^-)^2}\right)(s_b^+ + s_b^-)\right]. \qquad (11.23)$$

Moreover, we proved in 11.3 that:

$$(s_b^+ + s_b^-)^2 \stackrel{(law)}{=} \frac{\pi^2}{4}\,\mathbf{T}_{(2)},$$

so that the quantity in (11.23) is equal to:

$$E\left[f\left(\frac{1}{(\pi^2/4)\,\mathbf{T}_{(2)}}\right)\sqrt{\frac{\pi}{2}}\,\mathbf{T}_{(2)}\right],$$

which is the right-hand side of (11.22).

11.5 Another probabilistic representation of the Riemann zeta function

Given the relations, discussed above, between the distributions of m_e and $T_{(2)}$, the identity in law:

$$h_e \overset{\text{(law)}}{=} 2\, m_e, \qquad \text{where } h_e := \int_0^1 \frac{ds}{e(s)} \qquad (11.24)$$

obviously provides us with another probabilistic representation of the Riemann zeta function.

It will be shown below that (11.24) is a consequence of the following

Theorem 11.5 (Jeulin [52])   Let $(\ell_e^a;\ a \geq 0)$ be the family of local times of $(e(s),\ s \leq 1)$, and define:

$$k(t) = \sup\{y \geq 0;\ \int_y^\infty dx\, \ell_e^x > t\}.$$

Then, the process $((1/2)\ell_e^{k(t)};\ t \leq 1)$ is a normalized Brownian excursion.

We now prove (11.24). We deduce from Theorem 11.5 that:

$$h_e \overset{\text{(law)}}{=} \int_0^1 \frac{dt}{(1/2)\, \ell_e^{k(t)}},$$

and the right-hand side of this identity in law is equal to $2\, m_e$, which is obtained by making the change of variables $y = k(t)$.

11.6 Some generalizations related to Bessel processes

In this paragraph, the sequence $\mathbb{N}^*$ of positive integers will be replaced by the sequence of the zeros of the Bessel function J_μ.

Another important change with previous paragraphs is that, instead of studying m_μ^2, or $\sigma_\mu + \sigma_\mu'$ as in paragraph 11.2, in connection with the Riemann zeta function, it will be shown in this paragraph that the "Bessel zeta function"

ζ^ν which will be considered now has some close relationship with the time spent below 1 by a certain Bessel process.

(11.6.1) "Zeta functions" and probability.

It may be fruitless, for our purpose, to define which properties a "zeta function" should satisfy, e.g.: an Euler-product representation, or a functional equation, or ...; instead, we simply associate to a sequence $\lambda_* = (\lambda_n ; n \geq 1)$ of strictly positive real numbers, the "zeta function":

$$\zeta_{\lambda_*}(s) = \sum_{n=1}^{\infty} \frac{1}{\lambda_n^s} \ , \quad s > 0 \ .$$

In the sequel, we shall assume that: $\zeta_{\lambda_*}(1) = \sum_{n=1}^{\infty} \frac{1}{\lambda_n} < \infty$. We then have the elementary

Proposition 11.1 *Define the probability density:*
$\theta_{\lambda_*}(t) = c_{\lambda_*} \sum_{n=1}^{\infty} e^{-\lambda_n t}$ *with* $c_{\lambda_*} = \dfrac{1}{\zeta_{\lambda_*}(1)}$. *Then, if X_{λ_*} is a random variable with distribution $\theta_{\lambda_*}(t)dt$, we have:*

$$\zeta_{\lambda_*}(s)\Gamma(s) = \zeta_{\lambda_*}(1)E\left[(X_{\lambda_*})^{s-1}\right] \ , \quad s > 0 \ . \tag{11.25}$$

PROOF: This is an immediate consequence of the equality:

$$\Gamma(s)\frac{1}{a^s} = \int_0^\infty dx \, x^{s-1} e^{-ax} \ , \quad a > 0 \ , \ s > 0. \qquad \square$$

(11.6.2) Some examples related to Bessel processes.

a)   In this paragraph, we associate to any $\nu > 0$, the sequence:

$$\nu^* = (j_{\nu-1,n}^2 ; n \geq 1) \tag{11.26}$$

where $(j_{\mu,n} ; n \geq 1)$ denotes the increasing sequence of the simple, positive, zeros of the Bessel function J_μ (see Watson [90], p. 498).

We shall write $\zeta^\nu(s)$ for $\zeta_{\nu^*}(s)$, and $\theta^\nu(t)$ for $\theta_{\nu^*}(t)$. The aim of this paragraph is to exhibit a random variable $X^\nu \equiv X_{\nu^*}$ which is distributed as $\theta^\nu(t)dt$.

The following series representation shall play an essential rôle:

$$\frac{1}{x}\frac{I_\nu}{I_{\nu-1}}(x) = 2\sum_{n=1}^{\infty}\frac{1}{x^2 + j^2_{\nu-1,n}}\ ,\qquad x > 0 \qquad\qquad (11.27)$$

(see Watson [90], p. 498).

Now, we may prove the following

Theorem 11.6 *1) Let $y > 0$, and P_y^ν the law of the Bessel process $(R_t, t \geq 0)$, with index ν, starting from y at time 0. Then, we have:*

$$E_y^\nu\left[\exp -\alpha\int_0^\infty du\ 1_{(R_u \leq y)}\right] = \frac{2\nu}{y\sqrt{2\alpha}}\frac{I_\nu}{I_{\nu-1}}(y\sqrt{2\alpha}) \qquad (11.28)$$

2) Consequently, under P_y^ν, the distribution of the random variable:

$$X_y = \int_0^\infty du\ 1_{(R_u \leq y)} \qquad is \qquad \frac{1}{2y^2}\theta^\nu\left(\frac{t}{2y^2}\right)dt\ ,$$

where: $\qquad\theta^\nu(t) = (4\nu)\sum_{n=1}^{\infty}e^{-j^2_{\nu-1,n}t}\ ,\qquad t \geq 0\qquad \left(since:\ \zeta^\nu(1) = \frac{1}{4\nu}\right)$

$$(11.29)$$

Corollary 11.6.1 *For any $y > 0$, a candidate for the variable X^ν is*

$$\frac{1}{2y^2}X_y \equiv \frac{1}{2y^2}\int_0^\infty du\ 1_{(R_u \leq y)},\ \ under\ P_y^\nu.$$

Consequently, the following probabilistic representation of ζ^ν holds:

$$\zeta^\nu(s)\Gamma(s) = \frac{\zeta^\nu(1)}{(2y^2)^{s-1}}E_y^\nu\left[\left(\int_0^\infty du\ 1_{(R_u \leq y)}\right)^{s-1}\right]\ ,\qquad with\ \zeta^\nu(1) = \frac{1}{4\nu}\ .$$

$$(11.30)$$

PROOF OF THEOREM 11.6:

1) It may now be easier to use the following notation:
$\left(R_y^{(\nu)}(u), u \geq 0\right)$ denotes the Bessel process with index ν, starting at y at

time 0. Then, we have seen, and proved, in Chapter 4, the Ciesielski-Taylor identities:

$$\int_0^\infty du \; 1_{(R_0^{(\nu)}(u) \le y)} \stackrel{\text{(law)}}{=} T_y(R_0^{(\nu-1)})$$

Hence, with the help of this remark, and of the strong Markov property, we obtain:

$$E_y^\nu \left[\exp\left(-\alpha \int_0^\infty du \; 1_{(R_u \le y)} \right) \right] = \frac{E_0^{\nu-1}(\exp -\alpha T_y)}{E_0^\nu(\exp -\alpha T_y)}$$

and, to deduce formula (11.28), it suffices to use the following identity:

$$E_0^\mu(\exp -\alpha T_y) = \frac{(y\sqrt{2\alpha})^\mu}{2^\mu \Gamma(\mu+1) I_\mu(y\sqrt{2\alpha})} \quad , \tag{11.31}$$

for $\mu = \nu$, and $\mu = \nu - 1$ (see Kent [56], for example).

2) The proof of the second statement of the proposition now follows immediately from formulae (11.28) and (11.27). □

We now recall (see Chapter 6, in particular) that, if $(B_t, t \ge 0)$ denotes Brownian motion starting from 0, then $(\exp(B_t + \nu t); t \ge 0)$ may be represented as:

$$\exp(B_t + \nu t) = R^{(\nu)} \left(\int_0^t du \exp 2(B_u + \nu u) \right) \quad , \tag{11.32}$$

where $(R^{(\nu)}(t), t \ge 0)$ denotes here the Bessel process with index ν, starting from 1 at time 0. Hence, time-changing $R^{(\nu)}$ into $(\exp(B_t + \nu t), t \ge 0)$ with the help of formula (11.32), we obtain the following representation of $\zeta^\nu(s)$.

Corollary 11.6.2 Let $(B_t, t \ge 0)$ be a real valued Brownian motion starting from 0. Then, we have, for any $\nu > 0$:

$$\zeta^\nu(s) \Gamma(s) = \frac{\zeta^\nu(1)}{2^{s-1}} E \left[\left(\int_0^\infty du \exp 2(B_u + \nu u) 1_{(B_u + \nu u \le 0)} \right)^{s-1} \right] \tag{11.33}$$

(11.6.3) **The particular case** $\nu = \frac{3}{2}$.

We then have: $\nu - 1 = \frac{1}{2}$, and we are interested, from the definition of ν^* given in (11.26), in the sequence of positive zeros of

$$J_{\frac{1}{2}}(z) = \left(\frac{2}{\pi z}\right)^{1/2} \sin(z) \ .$$

Therefore, we have:

$$j_{1/2,n} = n\pi$$

Consequently, in the particular case $\nu = 3/2$, we may now write down the main result contained in Theorem 11.6 and its Corollaries, in the following form

Proposition 11.2 *We simply write* $\zeta_R(s) = \sum_{n=1}^{\infty} \frac{1}{n^s}$. *Then, we have*

$$3 \cdot 2^{s/2} \frac{\Gamma\left(\frac{s}{2}\right)}{\pi^s} \zeta_R(s) = E_1^{3/2}\left(\left(\int_0^{\infty} du\, 1_{(R_u \leq 1)}\right)^{\frac{s}{2}-1}\right)$$

(11.34)

$$= E\left[\left(\int_0^{\infty} dt\, \exp(2B_t + 3t) 1_{(2B_t + 3t \leq 0)}\right)^{\frac{s}{2}-1}\right]$$

11.7 Some relations between X^{ν} and $\Sigma^{\nu-1} \equiv \sigma_{\nu-1} + \sigma'_{\nu-1}$

(11.7.1) We begin with the most important case $\nu = \frac{3}{2}$, for which we simply write X for X^{ν} and Σ for $\Sigma^{\nu-1}$. Recall that, at the beginning of this Chapter, we used $\mathbf{T}_{(2)}$ as a notation for Σ, which now becomes more convenient.

Theorem 11.7 *Let* $X = \displaystyle\int_0^{\infty} ds\, 1_{(R_s^{(5)} \leq 1)}$, *where* $(R_s^{(5)}, s \geq 0)$ *denotes the Bessel process with dimension 5 (or index 3/2), starting from 1. Moreover, define:* $\Sigma \overset{(law)}{=} \sigma + \sigma'$, *where* σ *and* σ' *are two independent copies of the first hitting time of 1 by* $BES_0^{(3)}$.

Consider $\tilde{\Sigma}$, a random variable[3] which satisfies:

for every Borel function $f : \mathbb{R}_+ \to \mathbb{R}_+$, $E\left[f(\tilde{\Sigma})\right] = \frac{3}{2}E[f(\Sigma)\Sigma]$ (11.35)

Then, we have:

$$\text{a)} \qquad X \stackrel{\text{(law)}}{=} H\tilde{\Sigma} \qquad (11.36)$$

where H and $\tilde{\Sigma}$ are independent, and

$$P(H \in dh) = \left(\frac{1}{\sqrt{h}} - 1\right) dh \qquad (0 < h < 1)$$

or, equivalently:

$$H \stackrel{\text{(law)}}{=} UV^2 \stackrel{\text{(law)}}{=} (1 - \sqrt{U})^2,$$

where U and V denote two independent uniform r.v's;

$$\text{b)} \qquad \tilde{\Sigma} \stackrel{\text{(law)}}{=} \Sigma + X \qquad (11.37)$$

where, on the right-hand side, Σ and X are assumed to be independent.

Remark: The identity in law: $1 - \sqrt{U} \stackrel{\text{(law)}}{=} V\sqrt{U}$ which appears at the end of point a) above is a particular case of the identity in law between beta variables:

$$Z_{a,b+c} \stackrel{\text{(law)}}{=} Z_{a,b} Z_{a+b,c} \qquad \text{(see paragraph (8.1))}$$

with, here: $a = b = c = 1$

PROOF: a) Both identities in law (11.36) and (11.37) may be deduced from the explicit knowledge of the Laplace transforms of X and $\tilde{\Sigma}$, which are given by:

$$E\left[\exp(-\alpha X)\right] = 3\frac{\sqrt{2\alpha}\coth(\sqrt{2\alpha}) - 1}{2\alpha} \qquad (11.38)$$

(this is a particular case of formula (11.28), and

$$E\left[\exp(-\alpha\tilde{\Sigma})\right] = 3\frac{\sqrt{2\alpha}\coth(\sqrt{2\alpha}) - 1}{\text{sh}^2(\sqrt{2\alpha})} \qquad (11.39)$$

The identity in law (11.37) follows immediately from (11.38) and (11.39).

b) It may be interesting to give another proof of the identity in law (11.36). This second proof, which is in fact how the identity (11.36) was discovered, is obtained by comparing formula (11.34) with the definition of the function

---

[•] That is: $\tilde{\Sigma}$ is obtained by size-biased sampling of Σ.

$\xi(s)$, or rather with formula (11.3). By doing so, we obtain:

$$E[X^{\frac{s}{2}-1}] = \frac{3}{s(s-1)} E\left[\left(\frac{2}{\pi}\mathcal{N}\right)^{s/2}\right],$$

and, changing s into: $2k + 2$, we get:

$$E[X^k] = \frac{1}{(k+1)(2k+1)} \left(\frac{3}{2} E[\Sigma^{k+1}]\right) \quad (k \geq 0).$$

Now, we remark that

$$E[H^k] = E[U^k]\, E[V^{2k}] \equiv \frac{1}{(k+1)(2k+1)},$$

so that

$$E[X^k] = E[(H\tilde{\Sigma})^k] \quad (k \geq 0).$$

which implies (11.36). $\square$

Corollary 11.7.1 *(We use the same notations as in Theorem 11.7.)*

a) *The random variable Σ satisfies the identity in law*

$$\tilde{\Sigma} \overset{(\text{law})}{=} \Sigma + H\tilde{\Sigma}_1 \tag{11.40}$$

where, on the right-hand side, $\tilde{\Sigma}_1$ is independent of the pair (Σ, H), and is distributed as $\tilde{\Sigma}$.

b) *Equivalently, the function $g(\lambda) := E[\exp(-\lambda\Sigma)] \equiv \left(\dfrac{\sqrt{2\lambda}}{\text{sh}(\sqrt{2\lambda})}\right)^2$ satisfies:*

$$-\sqrt{\lambda}\,\frac{g'(\lambda)}{g(\lambda)} = \frac{1}{2}\int_0^\lambda \frac{dx}{x^{3/2}}\,(1 - g(x)). \tag{11.41}$$

PROOF: The identity (11.40) follows immediately from (11.36) and (11.37). We then deduce from (11.40) the identity

$$g'(\lambda) = g(\lambda) \int_0^1 dh \left(\frac{1}{\sqrt{h}} - 1\right) g'(h\lambda),$$

from which (11.41) follows, using integration by parts. $\square$

(11.7.2) We now present an extension for any ν of the identity in law (11.37).

Proposition 11.3 *Let* $X^\nu = \int_0^\infty ds\, 1_{(R^\nu_s \le 1)}$, *where* $(R^\nu_s, s \ge 0)$ *denotes the Bessel process with index* ν, *starting from 1, and define* $\Sigma^{\nu-1} = \sigma_{\nu-1} + \sigma'_{\nu-1}$, *where* $\sigma_{\nu-1}$ *and* $\sigma'_{\nu-1}$ *are two independent copies of the first hitting time of 1 by* $\mathrm{BES}_0^{\nu-1}$, *the Bessel process with index* $\nu - 1$ *starting from 0.*

Consider finally $\tilde{\Sigma}^{\nu-1}$, *a random variable which satisfies:*

for every Borel function $f : \mathbb{R}_+ \to \mathbb{R}_+, E\left[f(\tilde{\Sigma}^{\nu-1})\right] = \nu E\left[f(\Sigma^{\nu-1})\Sigma^{\nu-1}\right]$

Then, we have

$$\tilde{\Sigma}^{\nu-1} \overset{\text{(law)}}{=} \Sigma^{\nu-1} + X^\nu \qquad (11.42)$$

where the random variables on the right-hand side are assumed to be independent.

PROOF: From formula (11.31), we deduce:

$$E\left[\exp\left(-\lambda\Sigma^{\nu-1}\right)\right] = \left(\frac{(\sqrt{2\lambda})^{\nu-1}}{2^{\nu-1}\Gamma(\nu)I_{\nu-1}\left(\sqrt{2\lambda}\right)}\right)^2,$$

so that, taking derivatives with respect to λ on both sides, we obtain:

$$E\left[\Sigma^{\nu-1}\exp\left(-\lambda\Sigma^{\nu-1}\right)\right] = \left(\frac{x^{\nu-1}}{2^{\nu-1}\Gamma(\nu)I_{\nu-1}(x)}\right)^2 \left(\frac{2}{x}\frac{I_\nu}{I_{\nu-1}}(x)\right) \qquad (11.43)$$

where $x = \sqrt{2\lambda}$, and we have used the recurrence formula:

$$(\nu - 1)I_{\nu-1}(x) - xI'_{\nu-1}(x) = -xI_\nu(x) \ .$$

It now suffices to multiply both sides of (11.43) by ν and to use formula (11.28) to conclude. $\square$

Remark: The comparison of Theorem 11.7 and Proposition 11.3 suggests several questions, two of which are:

(i) is there an extension of the identity in law (11.36) for any ν, in the form:
$X^\nu \overset{\text{(law)}}{=} H_\nu \tilde{\Sigma}^{\nu-1}$, for some variable H_ν, which would be independent of $\tilde{\Sigma}^{\nu-1}$?

(ii) is there any relation between the functional equation for ζ and the identity in law (11.40), or equivalently (11.41)?

11.8 $\zeta^\nu(s)$ as a function of ν

In this paragraph, we show that the dependency in ν of the function $\zeta^\nu(s)$ may be understood as a consequence of the following Girsanov type relationship between the probability measures P_y^ν.

Theorem 11.8 *Let $y > 0$. On the canonical space $\Omega = C(\mathbb{R}_+, \mathbb{R}_+)$, we define $R_t(\omega) = \omega(t)$ ($t \geq 0$), and $\mathcal{L}_y(\omega) = \sup\{t \geq 0 : R_t(\omega) = y\}$. Then, as $\nu > 0$ varies, the measures $P_y^\nu\big|_{\mathcal{F}_{\mathcal{L}_y}}$ are all mutually absolutely continuous. More precisely, there exists a σ-finite measure $\mathbf{M}_y$ on $(\Omega, \mathcal{F}_{\mathcal{L}_y})$ such that, for every variable $Z \geq 0$, which is $\mathcal{F}_{\mathcal{L}_y}$ measurable, and every $\nu > 0$, we have:*

$$\mathbf{M}_y(Z) = \frac{1}{\nu} E_y^\nu \left[Z \exp \left(\frac{\nu^2}{2} \int_0^{\mathcal{L}_y} \frac{du}{R_u^2} \right) \right] . \tag{11.44}$$

PROOF: We consider the right-hand side of (11.44), and we disintegrate P_y^ν with respect to the law of $\mathcal{L}_y$. We obtain:

$$\frac{1}{\nu} E_y^\nu \left[Z \exp \left(\frac{\nu^2}{2} \int_0^{\mathcal{L}_y} \frac{du}{R_u^2} \right) \right]$$

$$= \frac{1}{\nu} \int P_y^\nu(\mathcal{L}_y \in dt) E_y^\nu \left[Z \exp \left(\frac{\nu^2}{2} \int_0^{\mathcal{L}_y} \frac{du}{R_u^2} \right) \mid \mathcal{L}_y = t \right] .$$

Now, it is well-known that conditioning with respect to $\mathcal{L}_y = t$ amounts to condition with respect to $R_t = y$ (see, for example, Revuz-Yor [81], Exercise (1.16), p.378) or Fitzsimmons-Pitman-Yor [41]); therefore, we have:

$$E_y^\nu \left[Z \exp \left(\frac{\nu^2}{2} \int_0^{\mathcal{L}_y} \frac{du}{R_u^2} \right) \mid \mathcal{L}_y = t \right] = E_y^\nu \left[Z \exp \left(\frac{\nu^2}{2} \int_0^{t} \frac{du}{R_u^2} \right) \mid R_t = y \right]$$

$$\tag{11.45}$$

Next, we use the absolute continuity relationship between P_y^ν and P_y^0:

$$P_y^\nu\big|_{\mathcal{F}_t} = \left(\frac{R_t}{y} \right)^\nu \exp \left(-\frac{\nu^2}{2} \int_0^{t} \frac{du}{R_u^2} \right) \cdot P_y^0\big|_{\mathcal{F}_t} ,$$

so that the expression in (11.45) is in fact equal to:

$$\frac{p_t^0(y,y)}{p_t^\nu(y,y)} E_y^0[Z \mid R_t = y] \ ,$$

where $\{p_t^\nu(x,y)\}$ is the family of densities of the semigroup $P_t^\nu(x; dy) \equiv p_t^\nu(x,y)dy$ associated to $\{P_x^\nu\}$.

Hence, the first expression we considered in the proof is equal to:

$$\frac{1}{\nu} E_y^\nu \left[Z \exp \left(\frac{\nu^2}{2} \int_0^{\mathcal{L}_y} \frac{du}{R_u^2} \right) \right] = \int_0^\infty \frac{P_y^\nu(\mathcal{L}_y \in dt)}{\nu p_t^\nu(y,y)} p_t^0(y,y) E_y^0[Z \mid R_t = y] \ .$$
(11.46)

However, it is known that:

$$P_y^\nu(\mathcal{L}_y \in dt) = \nu p_t^\nu(y,y)dt \qquad \text{(see Pitman-Yor [72])}$$

and finally, the expression in (11.46), which is equal to:

$$\int_0^\infty dt\, p_t^0(y,y) E_y^0[Z \mid R_t = y]$$

does not depend on ν.

Corollary 11.8.1 *1) Let $\tilde{\theta}^0(t)dt$ be the distribution of X_1 under the σ-finite measure $\mathbf{M}_1$. Then, the distribution of X_y under $\mathbf{M}_y$ is $\tilde{\theta}^0\left(\dfrac{t}{y^2}\right)\dfrac{dt}{y^2}$.*

2) For every $y > 0$, and $t > 0$, we have:

$$\frac{2}{\tilde{\theta}^0\left(\frac{t}{y^2}\right)} \sum_{n=1}^\infty e^{-(j_{\nu-1,n}^2)\left(\frac{t}{2y^2}\right)} = \mathbf{M}_y \left(\exp -\frac{\nu^2}{2} \int_0^{\mathcal{L}_y} \frac{du}{R_u^2} \mid X_y = t \right). \quad (11.47)$$

3) For every $\nu > 0$, we have:

$$\zeta^\nu(s)\Gamma(s) = \frac{1}{4} \frac{1}{(2y^2)^{s-1}} \mathbf{M}_y \left(\left(\int_0^\infty du\, 1_{(R_u \le y)} \right)^{s-1} \exp \left(-\frac{\nu^2}{2} \int_0^{\mathcal{L}_y} \frac{du}{R_u^2} \right) \right).$$
(11.48)

Consequently, the left-hand side of (11.47), i.e.: the "theta-function of index ν" and the left-hand side of (11.48), i.e.: the "zeta function of index ν" are Laplace transforms in $\left(\frac{\nu^2}{2}\right)$.

The last statement of the previous Corollary is confirmed by the explicit formulae found in Watson ([90], p. 502) for $\zeta^\nu(n)$, for n a small integer (Watson uses the notation $\sigma_{\nu-1}^{(s)}$ instead of our notation $\zeta^\nu(s)$).

In the following formulae, the function: $\nu \to \zeta^{\sqrt{\nu}}(n)$ appears to be a completely monotonic function of ν, as a sum (with positive coefficients) or a product of completely monotonic functions. Here are these formulae:

$$\zeta^{\sqrt{\nu}}(1) = \frac{1}{2^2\sqrt{\nu}} \qquad \zeta^{\sqrt{\nu}}(3) = \frac{1}{2^5\nu^{3/2}(\sqrt{\nu}+1)(\sqrt{\nu}+2)} \qquad (11.49)$$

$$\zeta^{\sqrt{\nu}}(2) = \frac{1}{2^4\nu(\sqrt{\nu}+1)} \qquad \zeta^{\sqrt{\nu}}(4) = \frac{5\sqrt{\nu}+6}{2^8\nu^2(\sqrt{\nu}+1)^2(\sqrt{\nu}+2)(\sqrt{\nu}+3)}$$

Comments on Chapter 11

The origin of this chapter is found in Biane-Yor [17]. We also recommend the more developed discussion in Biane [15]. D. Williams [93] presents a closely related discussion. Smith-Diaconis [84] start from the standard random walk before passing to the Brownian limit to obtain the functional equation (11.1).

A detailed discussion of the agreement formula (11.9) is found in Pitman-Yor [78].

References

1. J. Azéma and M. Yor. Sur les zéros des martingales continues. In *Séminaire de Probabilités, XXVI*, volume 1526 of *Lecture Notes in Math.*, pages 248–306. Springer, Berlin, 1992.
2. M. Barlow, J. Pitman, and M. Yor. Une extension multidimensionnelle de la loi de l'arc sinus. In *Séminaire de Probabilités, XXIII*, volume 1372 of *Lecture Notes in Math.*, pages 294–314. Springer, Berlin, 1989.
3. C. Bélisle. Windings of random walks. *Ann. Probab.*, 17(4):1377–1402, 1989.
4. C. Bélisle and J. Faraway. Winding angle and maximum winding angle of the two-dimensional random walk. *J. Appl. Probab.*, 28(4):717–726, 1991.
5. M. Berger and P. Roberts. On the winding number problem with finite steps. *Adv. in Appl. Probab.*, 20(2):261–274, 1988.
6. R. Berthuet. Étude de processus généralisant l'aire de Lévy. *Probab. Theory Related Fields*, 73(3):463–480, 1986.
7. J. Bertoin. Complements on the Hilbert transform and the fractional derivative of Brownian local times. *J. Math. Kyoto Univ.*, 30(4):651–670, 1990.
8. J. Bertoin. Excursions of a BES. (d) and its drift term $(0 < d < 1)$. *Probab. Theory Related Fields*, 84(2):231–250, 1990.
9. J. Bertoin. On the Hilbert transform of the local times of a Lévy process. *Bull. Sci. Math.*, 119(2):147–156, 1995.
10. J. Bertoin. *Lévy processes*, volume 121 of *Cambridge Tracts in Mathematics*. Cambridge University Press, Cambridge, 1996.
11. J. Bertoin and J. Pitman. Path transformations connecting Brownian bridge, excursion and meander. *Bull. Sci. Math.*, 118(2):147–166, 1994.
12. P. Biane. Comparaison entre temps d'atteinte et temps de séjour de certaines diffusions réelles. In *Séminaire de probabilités, XIX, 1983/84*, volume 1123 of *Lecture Notes in Math.*, pages 291–296. Springer, Berlin, 1985.
13. P. Biane. Sur un calcul de F. Knight. In *Séminaire de Probabilités, XXII*, volume 1321 of *Lecture Notes in Math.*, pages 190–196. Springer, Berlin, 1988.
14. P. Biane. Decomposition of brownian trajectories and some applications. Notes from lectures given at the Probability Winter School of Wuhan, China, Fall 1990.
15. P. Biane. La fonction zêta de Riemann et les probabilités. In *La fonction zêta*, pages 165–193. Ed. Éc. Polytech., Palaiseau, 2003.
16. P. Biane, J.-F. Le Gall, and M. Yor. Un processus qui ressemble au pont brownien. In *Séminaire de Probabilités, XXI*, volume 1247 of *Lecture Notes in Math.*, pages 270–275. Springer, Berlin, 1987.
17. P. Biane and M. Yor. Valeurs principales associées aux temps locaux browniens. *Bull. Sci. Math. (2)*, 111(1):23–101, 1987.

18. P. Biane and M. Yor. Quelques précisions sur le méandre brownien. *Bull. Sci. Math. (2)*, 112(1):101–109, 1988.

19. P. Biane and M. Yor. Sur la loi des temps locaux browniens pris en un temps exponentiel. In *Séminaire de Probabilités, XXII*, volume 1321 of *Lecture Notes in Math.*, pages 454–466. Springer, Berlin, 1988.

20. N. Bingham and R. Doney. On higher-dimensional analogues of the arc-sine law. *J. Appl. Probab.*, 25(1):120–131, 1988.

21. A. N. Borodin. Brownian local time. *Uspekhi Mat. Nauk*, 44(2(266)):7–48, 1989.

22. O. Brockhaus. The Martin boundary of the Brownian sheet. In *Stochastic partial differential equations (Edinburgh, 1994)*, volume 216 of *London Math. Soc. Lecture Note Ser.*, pages 22–30. Cambridge Univ. Press, Cambridge, 1995.

23. E. A. Carlen. The pathwise description of quantum scattering in stochastic mechanics. In *Stochastic processes in classical and quantum systems (Ascona, 1985)*, volume 262 of *Lecture Notes in Phys.*, pages 139–147. Springer, Berlin, 1986.

24. P. Carmona, F. Petit, and M. Yor. Sur les fonctionnelles exponentielles de certains processus de Lévy. *Stochastics Stochastics Rep.*, 47(1-2):71–101, 1994.

25. P. Carmona, F. Petit, and M. Yor. Beta-gamma random variables and intertwining relations between certain Markov processes. *Rev. Mat. Iberoamericana*, 14(2):311–367, 1998.

26. T. Chan, D. S. Dean, K. M. Jansons, and L. C. G. Rogers. On polymer conformations in elongational flows. *Comm. Math. Phys.*, 160(2):239–257, 1994.

27. K. L. Chung. Excursions in Brownian motion. *Ark. Mat.*, 14(2):155–177, 1976.

28. B. Davis. Brownian motion and analytic functions. *Ann. Probab.*, 7(6):913–932, 1979.

29. C. Dellacherie, P.-A. Meyer, and M. Yor. Sur certaines propriétés des espaces de Banach H^* et BMO. In *Séminaire de Probabilités, XII (Univ. Strasbourg, Strasbourg, 1976/1977)*, volume 649 of *Lecture Notes in Math.*, pages 98–113. Springer, Berlin, 1978.

30. C. Donati-Martin. Transformation de Fourier et temps d'occupation browniens. *Probab. Theory Related Fields*, 88(2):137–166, 1991.

31. C. Donati-Martin, S. Song, and M. Yor. Symmetric stable processes, Fubini's theorem, and some extensions of the Ciesielski-Taylor identities in law. *Stochastics Stochastics Rep.*, 50(1-2):1–33, 1994.

32. C. Donati-Martin and M. Yor. Mouvement brownien et inégalité de Hardy dans L^*. In *Séminaire de Probabilités, XXIII*, volume 1372 of *Lecture Notes in Math.*, pages 315–323. Springer, Berlin, 1989.

33. C. Donati-Martin and M. Yor. Fubini's theorem for double Wiener integrals and the variance of the Brownian path. *Ann. Inst. H. Poincaré Probab. Statist.*, 27(2):181–200, 1991.

34. C. Donati-Martin and M. Yor. Some Brownian functionals and their laws. *Ann. Probab.*, 25(3):1011–1058, 1997.

35. L. E. Dubins and M. Smorodinsky. The modified, discrete, Lévy-transformation is Bernoulli. In *Séminaire de Probabilités, XXVI*, volume 1526 of *Lecture Notes in Math.*, pages 157–161. Springer, Berlin, 1992.

36. B. Duplantier. Areas of planar Brownian curves. *J. Phys. A*, 22(15):3033–3048, 1989.

37. R. Durrett. A new proof of Spitzer's result on the winding of two-dimensional Brownian motion. *Ann. Probab.*, 10(1):244–246, 1982.

38. E. Dynkin. Some limit theorems for sums of independent random variables with infinite mathematical expectations. In *Select. Transl. Math. Statist. and Probability, Vol. 1*, pages 171–189. Inst. Math. Statist. and Amer. Math. Soc., Providence, R.I., 1961.

39. N. Eisenbaum. Un théorème de Ray-Knight lié au supremum des temps locaux browniens. *Probab. Theory Related Fields*, 87(1):79–95, 1990.

40. P. Fitzsimmons and R. Getoor. On the distribution of the Hilbert transform of the local time of a symmetric Lévy process. *Ann. Probab.*, 20(3):1484–1497, 1992.

41. P. Fitzsimmons, J. Pitman, and M. Yor. Markovian bridges: construction, Palm interpretation, and splicing. In *Seminar on Stochastic Processes, 1992 (Seattle, WA, 1992)*, volume 33 of *Progr. Probab.*, pages 101–134. Birkhäuser Boston, Boston, MA, 1993.

42. A. Földes and P. Révész. On hardly visited points of the Brownian motion. *Probab. Theory Related Fields*, 91(1):71–80, 1992.

43. H. Föllmer. Martin boundaries on Wiener space. In *Diffusion processes and related problems in analysis, Vol. I (Evanston, IL, 1989)*, volume 22 of *Progr. Probab.*, pages 3–16. Birkhäuser Boston, Boston, MA, 1990.

44. G. J. Foschini and L. A. Shepp. Closed form characteristic functions for certain random variables related to Brownian motion. In *Stochastic analysis*, pages 169–187. Academic Press, Boston, MA, 1991. Liber amicorum for Moshe Zakai.

45. M. Fukushima. A decomposition of additive functionals of finite energy. *Nagoya Math. J.*, 74:137–168, 1979.

46. H. Geman and M. Yor. Quelques relations entre processus de Bessel, options asiatiques et fonctions confluentes hypergéométriques. *C. R. Acad. Sci. Paris Sér. I Math.*, 314(6):471–474, 1992.

47. H. Geman and M. Yor. Bessel processes, asian options and perpetuities. *Math. Finance*, 3 (4):349–375, 1993.

48. P. Hartman and G. Watson. "Normal" distribution functions on spheres and the modified Bessel functions. *Ann. Probability*, 2:593–607, 1974.

49. J.-P. Imhof. Density factorizations for Brownian motion, meander and the three-dimensional Bessel process, and applications. *J. Appl. Probab.*, 21(3):500–510, 1984.

50. K. Itô and H. McKean. *Diffusion processes and their sample paths*. Springer-Verlag, Berlin, 1974. Second printing, corrected, Die Grundlehren der mathematischen Wissenschaften, Band 125.

51. T. Jeulin. *Semi-martingales et grossissement d'une filtration*, volume 833 of *Lecture Notes in Mathematics*. Springer, Berlin, 1980.

52. T. Jeulin. Application de la théorie du grossissement à l'étude des temps locaux browniens,. In *Grossissement de filtrations: Exemples et applications*, volume 1118 of *Lecture Notes in Math.*, pages 197–304. Springer, Berlin, 1985.

53. T. Jeulin and M. Yor. Inégalité de Hardy, semimartingales, et faux-amis. In *Séminaire de Probabilités, XIII (Univ. Strasbourg, Strasbourg, 1977/78)*, volume 721 of *Lecture Notes in Math.*, pages 332–359. Springer, Berlin, 1979.

54. T. Jeulin and M. Yor. Filtration des ponts browniens et équations différentielles stochastiques linéaires. In *Séminaire de Probabilités, XXIV, 1988/89*, volume 1426 of *Lecture Notes in Math.*, pages 227–265. Springer, Berlin, 1990.

55. Y. Kasahara and S. Kotani. On limit processes for a class of additive functionals of recurrent diffusion processes. *Z. Wahrsch. Verw. Gebiete*, 49(2):133–153, 1979.

56. J. Kent. Some probabilistic properties of Bessel functions. *Ann. Probab.*, 6(5):760–770, 1978.

57. F. B. Knight. Random walks and a sojourn density process of Brownian motion. *Trans. Amer. Math. Soc.*, 109:56–86, 1963.

58. F. B. Knight. Inverse local times, positive sojourns, and maxima for Brownian motion. *Astérisque*, (157-158):233–247, 1988. Colloque Paul Lévy sur les Processus Stochastiques (Palaiseau, 1987).

59. J.-F. Le Gall. Mouvement brownien, cônes et processus stables. *Probab. Theory Related Fields*, 76(4):587–627, 1987.

60. J.-F. Le Gall and M. Yor. Excursions browniennes et carrés de processus de Bessel. *C. R. Acad. Sci. Paris Sér. I Math.*, 303(3):73–76, 1986.

61. J.-F. Le Gall and M. Yor. Étude asymptotique des enlacements du mouvement brownien autour des droites de l'espace. *Probab. Theory Related Fields*, 74(4):617–635, 1987.

62. J.-F. Le Gall and M. Yor. Enlacements du mouvement brownien autour des courbes de l'espace. *Trans. Amer. Math. Soc.*, 317(2):687–722, 1990.

63. N. N. Lebedev. *Special functions and their applications*. Dover Publications Inc., New York, 1972. Revised edition, translated from the Russian and edited by Richard A. Silverman, Unabridged and corrected republication.

64. P. Lévy. Sur certains processus stochastiques homogènes. *Compositio Math.*, 7:283–339, 1939.

65. P. Messulam and M. Yor. On D. Williams' "pinching method" and some applications. *J. London Math. Soc. (2)*, 26(2):348–364, 1982.

66. F. Mezzadri and N. C. Snaith, editors. *Recent perspectives in random matrix theory and number theory*, volume 322 of *London Mathematical Society Lecture Note Series*. Cambridge University Press, Cambridge, 2005.

67. E. Molčanov and S. Ostrovskiĭ. Symmetric stable processes as traces of degenerate diffusion processes. *Teor. Verojatnost. i Primenen.*, 14:127–130, 1969.

68. E. Perkins. Local time is a semimartingale. *Z. Wahrsch. Verw. Gebiete*, 60(1):79–117, 1982.

69. M. Perman, J. Pitman, and M. Yor. Size-biased sampling of Poisson point processes and excursions. *Probab. Theory Related Fields*, 92(1):21–39, 1992.

70. F. Petit. *Sur le temps passé par le mouvement brownien au-dessus d'un multiple de son supremum, et quelques extensions de la loi de l'arc sinus*. PhD thesis, Université Paris VII, February 1992.

71. J. Pitman. One-dimensional Brownian motion and the three-dimensional Bessel process. *Advances in Appl. Probability*, 7(3):511–526, 1975.

72. J. Pitman and M. Yor. Bessel processes and infinitely divisible laws. In *Stochastic integrals (Proc. Sympos., Univ. Durham, Durham, 1980)*, volume 851 of *Lecture Notes in Math.*, pages 285–370. Springer, Berlin, 1981.

73. J. Pitman and M. Yor. A decomposition of Bessel bridges. *Z. Wahrsch. Verw. Gebiete*, 59(4):425–457, 1982.

74. J. Pitman and M. Yor. Sur une décomposition des ponts de Bessel. In *Functional analysis in Markov processes (Katata/Kyoto, 1981)*, volume 923 of *Lecture Notes in Math.*, pages 276–285. Springer, Berlin, 1982.

75. J. Pitman and M. Yor. Asymptotic laws of planar Brownian motion. *Ann. Probab.*, 14(3):733–779, 1986.

76. J. Pitman and M. Yor. Further asymptotic laws of planar Brownian motion. *Ann. Probab.*, 17(3):965–1011, 1989.

77. J. Pitman and M. Yor. Arcsine laws and interval partitions derived from a stable subordinator. *Proc. London Math. Soc. (3)*, 65(2):326–356, 1992.

78. J. Pitman and M. Yor. Decomposition at the maximum for excursions and bridges of one-dimensional diffusions. In *Itô's stochastic calculus and probability theory*, pages 293–310. Springer, Tokyo, 1996.

79. J. Pitman and Marc Yor. Dilatations d'espace-temps, réarrangements des trajectoires browniennes, et quelques extensions d'une identité de Knight. *C. R. Acad. Sci. Paris Sér. I Math.*, 316(7):723–726, 1993.

80. D. Ray. Sojourn times of diffusion processes. *Illinois J. Math.*, 7:615–630, 1963.

81. D. Revuz and M. Yor. *Continuous martingales and Brownian motion*, volume 293 of *Grundlehren der Mathematischen Wissenschaften [Fundamental Principles of Mathematical Sciences]*. Springer-Verlag, Berlin, third edition, 1999.

82. J. Rudnick and Y. Hu. The winding angle distribution of an ordinary random walk. *J. Phys. A*, 20(13):4421–4438, 1987.

83. T. Shiga and S. Watanabe. Bessel diffusions as a one-parameter family of diffusion processes. *Z. Wahrscheinlichkeitstheorie und Verw. Gebiete*, 27:37–46, 1973.

84. L. Smith and P. Diaconis. Honest Bernoulli excursions. *J. Appl. Probab.*, 25(3):464–477, 1988.

85. F. Spitzer. Some theorems concerning 2-dimensional Brownian motion. *Trans. Amer. Math. Soc.*, 87:187–197, 1958.

86. A. Truman and D. Williams. A generalised arc-sine law and Nelson's stochastic mechanics of one-dimensional time-homogeneous diffusions. In *Diffusion processes and related problems in analysis, Vol. I (Evanston, IL, 1989)*, volume 22 of *Progr. Probab.*, pages 117–135. Birkhäuser Boston, Boston, MA, 1990.

87. A. Truman and D. Williams. Excursions and Itô calculus in Nelson's stochastic mechanics. In *Recent developments in quantum mechanics (Poiana Braşov, 1989)*, volume 12 of *Math. Phys. Stud.*, pages 49–83. Kluwer Acad. Publ., Dordrecht, 1991.

88. P. Vallois. Sur la loi conjointe du maximum et de l'inverse du temps local du mouvement brownien: application à un théorème de Knight. *Stochastics Stochastics Rep.*, 35(3):175–186, 1991.

89. W. Vervaat. A relation between Brownian bridge and Brownian excursion. *Ann. Probab.*, 7(1):143–149, 1979.

90. G. Watson. *A treatise on the theory of Bessel functions*. Cambridge Mathematical Library. Cambridge University Press, Cambridge, 1995. Reprint of the second (1944) edition.

91. M. L. Wenocur. Brownian motion with quadratic killing and some implications. *J. Appl. Probab.*, 23(4):893–903, 1986.

92. M. L. Wenocur. Ornstein-Uhlenbeck process with quadratic killing. *J. Appl. Probab.*, 27(3):707–712, 1990.

93. D. Williams. Brownian motion and the Riemann zeta-function. In *Disorder in physical systems*, Oxford Sci. Publ., pages 361–372. Oxford Univ. Press, New York, 1990.

94. T. Yamada. On the fractional derivative of Brownian local times. *J. Math. Kyoto Univ.*, 25(1):49–58, 1985.

95. T. Yamada. On some limit theorems for occupation times of one-dimensional Brownian motion and its continuous additive functionals locally of zero energy. *J. Math. Kyoto Univ.*, 26(2):309–322, 1986.

96. T. Yamada. Representations of continuous additive functionals of zero energy via convolution type transforms of Brownian local times and the Radon transform. *Stochastics Stochastics Rep.*, 48(1-2):1–15, 1994.

97. T. Yamada. Principal values of Brownian local times and their related topics. In *Itô's stochastic calculus and probability theory*, pages 413–422. Springer, Tokyo, 1996.

98. M. Yor. Loi de l'indice du lacet brownien, et distribution de Hartman-Watson. *Z. Wahrsch. Verw. Gebiete*, 53(1):71–95, 1980.

99. M. Yor. Une extension markovienne de l'algèbre des lois béta-gamma. *C. R. Acad. Sci. Paris Sér. I Math.*, 308(8):257–260, 1989.

100. M. Yor. Étude asymptotique des nombres de tours de plusieurs mouvements browniens complexes corrélés. In *Random walks, Brownian motion, and interacting particle systems*, volume 28 of *Progr. Probab.*, pages 441–455. Birkhäuser Boston, Boston, MA, 1991.

101. M. Yor. Une explication du théorème de Ciesielski-Taylor. *Ann. Inst. H. Poincaré Probab. Statist.*, 27(2):201–213, 1991.

102. M. Yor. On some exponential functionals of Brownian motion. *Adv. in Appl. Probab.*, 24(3):509–531, 1992.

103. M. Yor, editor. *Exponential functionals and principal values related to Brownian motion*. Biblioteca de la Revista Matemática Iberoamericana. [Library of the Revista Matemática Iberoamericana]. Revista Matemática Iberoamericana, Madrid, 1997. A collection of research papers.

104. M. Yor. *Exponential functionals of Brownian motion and related processes*. Springer Finance. Springer-Verlag, Berlin, 2001. With an introductory chapter by Hélyette Geman, Chapters 1, 3, 4, 8 translated from the French by Stephen S. Wilson.

105. M. Yor, M. Chesney, H. Geman, and M. Jeanblanc-Picqué. Some combinations of Asian, Parisian and barrier options. In *Mathematics of derivative securities (Cambridge, 1995)*, volume 15 of *Publ. Newton Inst.*, pages 61–87. Cambridge Univ. Press, Cambridge, 1997.

Further general references about Brownian Motion and Related Processes

1. J. Bertoin. *Random fragmentation and coagulation processes*, volume 102 of *Cambridge Studies in Advanced Mathematics*. Cambridge University Press, Cambridge, 2006.
2. A. N. Borodin and P. Salminen. *Handbook of Brownian motion—facts and formulae*. Probability and its Applications. Birkhäuser Verlag, Basel, second edition, 2002.
3. J. L. Doob. *Classical potential theory and its probabilistic counterpart*, volume 262 of *Grundlehren der Mathematischen Wissenschaften [Fundamental Principles of Mathematical Sciences]*. Springer-Verlag, New York, 1984.
4. R. Durrett. *Brownian motion and martingales in analysis*. Wadsworth Mathematics Series. Wadsworth International Group, Belmont, CA, 1984.
5. F. B. Knight. *Essentials of Brownian motion and diffusion*, volume 18 of *Mathematical Surveys*. American Mathematical Society, Providence, R.I., 1981.
6. G. F. Lawler. *Conformally invariant processes in the plane*, volume 114 of *Mathematical Surveys and Monographs*. American Mathematical Society, Providence, RI, 2005.
7. J.-F. Le Gall. Some properties of planar Brownian motion. In *École d'Été de Probabilités de Saint-Flour XX—1990*, volume 1527 of *Lecture Notes in Math.*, pages 111–235. Springer, Berlin, 1992.
8. M. B. Marcus and J. Rosen. *Markov processes, Gaussian processes, and local times*, volume 100 of *Cambridge Studies in Advanced Mathematics*. Cambridge University Press, Cambridge, 2006.
9. J. Pitman. *Combinatorial stochastic processes*, volume 1875 of *Lecture Notes in Mathematics*. Springer-Verlag, Berlin, 2006. Lectures from the 32nd Summer School on Probability Theory held in Saint-Flour, July 7–24, 2002, With a foreword by Jean Picard.
10. M. Rao. *Brownian motion and classical potential theory*. Matematisk Institut, Aarhus University, Aarhus, 1977. Lecture Notes Series, No. 47.
11. L. C. G. Rogers and D. Williams. *Diffusions, Markov processes, and martingales. Vol. 1*. Cambridge Mathematical Library. Cambridge University Press, Cambridge, 2000. Foundations, Reprint of the second (1994) edition.
12. L. C. G. Rogers and D. Williams. *Diffusions, Markov processes, and martingales. Vol. 2*. Cambridge Mathematical Library. Cambridge University Press, Cambridge, 2000. Itô calculus, Reprint of the second (1994) edition.
13. W. Werner. Random planar curves and Schramm-Loewner evolutions. In *Lectures on probability theory and statistics*, volume 1840 of *Lecture Notes in Math.*, pages 107–195. Springer, Berlin, 2004.

Universitext